P9-DFI-605

Praise for

Plowing with Pigs

From hogging stumps to outwitting weasels, Oscar and Karen Will know all the fun, sweat and tears needed to live off the land. Here, they generously share their hard-learned lessons. Newcomers to the world of homesteading have much to gain from reading this book.

— Shannon Hayes, author of *Radical Homemakers* and *Long Way on a Little*

This book contains a wonderful collection of ideas and advice that will be invaluable to anyone whose goal is to build a more satisfying, sustainable life on the land. Hank and Karen Will are highly skilled farmers/gardeners/cooks/innovators who will teach you how to blend the best of modern tools with old-time country skills. Their "you can do it" attitude provides abundant inspiration and their clever ideas will save any homesteader a great deal of money.

— Cheryl Long, Editor-in-Chief, *Mother Earth News*

This book is a resource of limitless value. The Wills describe a life in which personal effort and ingenuity create beauty, health and sustenance. Their practical advice and first-person anecdotes provide an ideal guide for anyone interested in self-reliance and a naturally wholesome lifestyle. You won't find a better model for a life lived well.

— Bryan Welch, Publisher & Editorial Director, *Mother Earth News, GRIT, Mother Earth Living*

Leave it to Karen and Hank Will to write a great introductory book for the aspiring back-to-the-lander. In *Plowing with Pigs* the Wills cover everything from raising farm animals to growing crops, and from building farmstead improvements to cooking with farm-fresh produce. The Wills not only describe the how-to's of a small-scale farm or homestead in a way that is easy to follow, but also share many real-life lessons from their own experiences.

— CAROL EKARIUS, author of *Storey's Guide to Poultry Breeds*

Whatever motivates your reach for greater self-sufficiency—an aspiration toward a more fulfilling life or the looming fractures in our economic systems—you will find abundant practical guidance in *Plowing with Pigs*. Out of a deep well of hard-won and hands-on experience, Hank and Karen Will offer an impressive array of strategies, grounded in a few principles both profound and simple: using cost-free assets available on the farm or homestead, using domesticated animals as partners—and using our heads.

—HARVEY USSERY, author of *The Small-Scale Poultry Flock*

Whether you're an established homesteader or just thinking of a homesteading lifestyle, grab a cup of tea or coffee and curl up with this book. Hank and Karen are the friendly neighbors with sage advice that homesteaders relied upon a century ago. In creating their own sustainable farm, they've discovered hundreds of practices that can make everyone's homestead run more smoothly. (I can't wait to create my own chicken moat!) *Plowing With Pigs* should be on the shelf of everyone who wants to live a more self-reliant lifestyle.

— DEBORAH NIEMANN, author of *Homegrown and Handmade* and *Ecothrifty*

Plowing *with* Pigs

PLOWING *with* PIGS

AND OTHER CREATIVE, LOW-BUDGET HOMESTEADING SOLUTIONS

Oscar H. Will III & Karen K. Will

Cover design by Diane McIntosh.
Boots image © istock (akpakp); Pig image © iStock (Tana26)

Printed in Canada. First printing January 2013.

Paperback ISBN: 978-0-86571-717-6
eISBN: 978-1-55092-523-4

Inquiries regarding requests to reprint all or part of *Plowing with Pigs* should be addressed to New Society Publishers at the address below.
To order directly from the publishers, please call toll-free (North America) 1-800-567-6772, or order online at www.newsociety.com

Any other inquiries can be directed by mail to:

New Society Publishers
P.O. Box 189, Gabriola Island, BC V0R 1X0, Canada
(250) 247-9737

New Society Publishers' mission is to publish books that contribute in fundamental ways to building an ecologically sustainable and just society, and to do so with the least possible impact on the environment, in a manner that models this vision. We are committed to doing this not just through education, but through action. The interior pages of our bound books are printed on Forest Stewardship Council®-registered acid-free paper that is **100% post-consumer recycled** (100% old growth forest-free), processed chlorine free, and printed with vegetable-based, low-VOC inks, with covers produced using FSC®-registered stock. New Society also works to reduce its carbon footprint, and purchases carbon offsets based on an annual audit to ensure a carbon neutral footprint. For further information, or to browse our full list of books and purchase securely, visit our website at: **www.newsociety.com**

LIBRARY AND ARCHIVES CANADA CATALOGUING IN PUBLICATION

Will, Oscar H., 1956-

Plowing with pigs : and other creative, low-budget homesteading solutions / Oscar H. Will and Karen K. Will.

Includes index.
ISBN 978-0-86571-717-6

1. Agriculture--Handbooks, manuals, etc. 2. Farm life--Handbooks, manuals, etc. 3. Self-reliant living. I. Will, Karen K. II. Title.

S501.2.W53 2013 640 C2012-907889-1

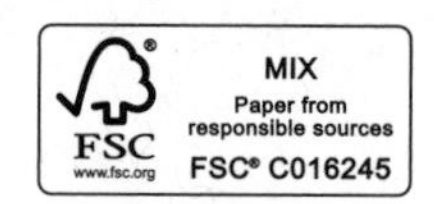

Contents

Books for Wiser Living recommended by *Mother Earth News*

Today, more than ever before, our society is seeking ways to live more conscientiously. To help bring you the very best inspiration and information about greener, more sustainable lifestyles, *Mother Earth News* is recommending select books from New Society Publishers. For more than 30 years, *Mother Earth News* has been North America's "Original Guide to Living Wisely," creating books and magazines for people with a passion for self-reliance and a desire to live in harmony with nature. Across the countryside and in our cities, New Society Publishers and *Mother Earth News* are leading the way to a wiser, more sustainable world. For more information, please visit MotherEarthNews.com.

Acknowledgments

It would be impossible to name all of the people who have inspired us along the road to this project — some of you know who you are, while others carried out their work through their writings. We would like to thank all of you for providing encouragement, mentoring, honest critique, and even constructive, yet gentle, criticism of our thought processes and approaches.

We'd especially like to thank friend, mentor, and Ogden Publications (*GRIT, Mother Earth News, Mother Earth Living,* etc.) founder and leader, Bryan Welch, for creating an environment that made it possible for us to thrive and hone our writing skills, and for offering us his complete support — even to the extent of providing expert sheep-handling advice on many occasions. Bryan provided the opportunity for us to speak about the topics covered in the book at the *Mother Earth News Fair* events, which led to a fortuitous encounter with Heather Nicholas at New Society Publishers.

Heather graciously sat through one of our presentations, waited until all of the lingering questions were answered, then introduced herself and asked whether we'd ever considered organizing our ideas into book form. Thank you, Heather, for encouraging us and leading us successfully through the book proposal process. Thanks also to the entire NSP crew. We've not worked with a more gracious bunch of book publishing folks.

Thanks also to our copy editor, Linda Glass. She helped this pair of picky magazine editors create a product that is clear, lean, and we hope more enjoyable to read. Linda also had some great ideas on organizing information within chapters — thanks.

We'd like to mention 21st-century homestead practitioners Wendy and Eric Slatt for inspirational discussion — mostly via email — and photos. Thanks also to Nathan Winters, the out-of-the-box farmer and thinker, and to Suzanne Whisnant Cox for so cheerfully and successfully taking the plowing with pigs concept and putting it into practice. And many thanks to our local homesteading friends who are always up for discussing alternative ways to accomplish things: Nathan Lindsay, Robin Mather, Jennifer Kongs, and Allan and Heather Scherger. All of you help to remind us that doing what we do as part of a community is more fun and more productive than going it alone.

Finally, we'd like to thank you, dear reader, for your interest in finding new and new/old ways to carve out a piece of the good life. Without you and your enthusiastic passion, there'd be no reason for this book.

Introduction

Fueled by a failing economy and an American Dream that will remain a mere shadow of its former consumption-based self, homesteaders everywhere are searching for that elusive good life. These 21st-century pioneers span the generations and are led by a passionately enthusiastic group of young folks who have nothing to lose monetarily or materially by eschewing mainstream mores that would hitch them to the corporate mill. And yet, the very fact that money is in relatively short supply makes it psychologically difficult for committed folks to achieve independence because they still feel the need to buy all the "stuff" that "they" say we need.

Sustainability, the buzzword of the 21st-century, takes on new meaning as corporate interests compete to own the concept. As they busy themselves jockeying to *look* sustainable, we feel it is important to ask ourselves how we should be responding to ever-dwindling energy resources. What does it really mean to be sustainable, and how can we achieve it? What should it mean at the level of the individual, a household, or a community?

We believe that there are lessons to be learned from the time when households and local economies were largely self-sustainable. If we learn those lessons, we may be able to deliver a world worth living in to future generations. We aren't suggesting a regression back to the dark ages, but we are enthusiastic about a future in which households once again produce much of their own food, energy, goods, and tools — and people make better use of readily available renewable resources. Imagine a society that values do-it-yourself skills, leisure that promotes intellectual development and creativity, and the making of one's own fun.

We decided to write this book because we're passionate about these topics. We believe that a fruitful future includes developing a society that can care for itself in a humane way. The major issues that we address relate to the widespread belief that you need all kinds of "stuff" to live the good life — that money is the only commodity worth working for. We believe that you can use your head, heart, and hands in lieu of cash. Plus, we want to reintroduce an almost forgotten resource: partnering with domesticated animals. Using animals to help us with our labors is a time-tested and effective means to self-sustainability. Domesticated animals have many purposes — only one of those purposes is food.

This is not just another homesteading book. We specifically do not rehash specific food preservation methods or animal husbandry basics. Instead, we suggest a paradigm shift in thinking about how to carve out a truly sustainable life in the 21st-century. You can't buy this life. You can only live it. We offer readers a number of entry points to this way of life. This book is about eschewing materialism and blind consumerism, doing more for yourself, and relying less on corporate machines as you search for true satisfaction.

Author-practitioners who have inspired us over the years include John Seymour, Joel Salatin, Shannon Hayes, Wendell Berry, Harvey Ussery, André Voisin, Carla Emery, Kelly Klober, Gene Logsdon, Aldo Leopold, William Woys Weaver, Carol Ekarius, Darina Allen, Sue Weaver, and a host of others. We hope to inspire you — should we fail, try reading these authors to keep your fires of passion stoked.

Section 1

Animals

Fowl, hogs, and small ruminants such as sheep and goats are hallmarks of the 21st-century's rural homestead. Flocks of chickens, at the very least, characterize urban and suburban versions of the same. Though sometimes raised as pets or for some specific commodity like fiber, eggs, or meat, domesticated farm animals of all kinds can empower you to do more with less — even if you have no intention of eating the animals. Chickens can rid your yard of ticks and other insect pests, sheep can mow the lawn, and goats will gladly rid your pastures and fence lines of poison ivy and woody brush. These animals provide not only rich, organic fertilizers to help build soil, they can also provide endless hours of entertainment.

Understanding a specific breed's or species' behavioral biology is the first step to finding ways to put animal allies to work. The goal is to simply allow them to do what comes naturally. Casting some conventional thinking to the wind is the second step; our ancestors knew to let animals work for them, but today's highly controlled industrial agriculture models don't have room for letting animals do their thing. The final step in getting animals to do some of your pesky work is to give it a try and have some fun with it.

In the following three chapters, we consider some of our ancestors' animal understanding — with a 21st-century twist.

CHAPTER 1

Free-Range Fowl

At the turn of the 20th-century, a common sight in backyards and farms throughout the country was the family cow, a small chicken flock, and even the occasional shoat being fattened on kitchen scraps. Over the years, as folks sought to prove they were sophisticated and not "fresh from the farm," municipal codes were changed. Residents were denied the right to keep animals that had provided a measure of self-sufficiency. So thorough was the campaign against chickens that many people now view *Gallus gallus* as the source of every potential sanitation and nuisance problem in a city. Code enforcement officers would have you believe that chickens by their very nature "attract vermin" (such as mice and rats), "draw flies," or literally "raise a stink." Plus chickens are noisy. Ha!

Truth be told, chickens don't attract mice and flies, but they do relish eating any that do show up. They'll eat every fly larva they find. And when chickens are kept in a proper coop or sufficient-sized yard, their manure is never sufficiently moist or concentrated to generate an offensive ammonia odor. And even if manure builds up below their perches, chickens will naturally scratch through that manure and the bedding, which keeps the decomposition process aerobic. We don't deny that roosters can be annoyingly loud in town, but the murmuring discussion among contented hens can hardly be called loud. And even when the girls announce the impending arrival of an egg, that clucking is not as loud or annoying as a barking dog — yet, dogs are allowed virtually everywhere. Not so the chicken.

We're pleased to report that the chicken has made a comeback! Twenty-first-century homesteaders living out where the pavement ends are adding flocks in droves, and people living in town and suburbia are demanding their right to keep chickens by working to get their municipal codes changed. In most cases, urban homesteaders are legally limited to the number (and in some cases, the type) of chickens they can keep, but the birds are reclaiming their place in North American backyards one town at a time. There are many reasons to factor chickens into your homesteading plans — even if you have no intention of eating any part of the chicken or its byproducts.

New Work for Old Chickens

Back in the day, fowl fanciers and farmstead owners all over the country kept chickens because they were beautiful, particularly suited to a specific region's environment, and for the services they could offer and/or products they could supply. Did you know that some fancy fowl were kept to supply the fashion and fishing-fly-tying industries with incredibly beautiful feathers, which were often harvested without ending the bird's life? Others were kept for the eggs or meat they could provide. And all the while, the birds kept their premises free of all manner of pests, including flies, ticks, grubs, caterpillars, and even mice and snakes

Fig. 1.1: *Chickens left to roam freely will till the soil in your perennial beds and keep insect pests at bay as well as add beauty to the landscaping.*
Credit: Oscar H. Will III

in some instances. Plus, chickens are an end-of-the-day entertainment that rivals the best Broadway show or blockbuster movie. You've heard the expression "sit and watch the chickens peck." For the homesteader, there may be nothing quite so soothing at the end of a fulfilling day of work than to sit, cool beverage in hand, and watch the chickens just do what chickens do.

So, what is it that they do? Well, if the chickens in question happen to be one of the small handful of over-bred industrial breeds, those poor animals will have few social skills and may grow so fast and so out of proportion that they break bones or die of heart attacks just eight weeks after hatching. Watching these chickens do their thing may be more depressing than relaxing or uplifting — especially if they're in a horrific factory-production setting — but that's not their fault. It's the fault of our morally bankrupt, so-called land-grant-university animal science departments, which — in collusion with the very industry profiting from the "research" — have determined that animals such as chickens are nothing more than cogs in a money-making machine. As such, laying hens may legally be crowded into small cages where they cannot scratch, cannot interact socially with one another, and cannot lay eggs in the privacy of a nest box or other "secret" place. On top of all of that, most of the top-half of their beaks have been cut off to keep the overcrowded animals from pecking one another.

Thankfully, a sufficient number of folks interested in animal husbandry eschewed the entire industrial poultry production model and have maintained many of the old chicken breeds and lines. Thus it is that some of those sturdy, older breeds are available today. Birds like the Jersey Giant will net you some eggs and grow to sufficient size to produce a fine table fowl. But more importantly, old breeds like the Jersey Giant thrive out-of-doors, and they will entertain you beautifully while performing tasks you'd rather not do and doing the work of agricultural poisons and synthetic chemicals you'd rather not use.

Fig. 1.2: *Lead your flock to fresh forage using one of their favorite treats such as a mixture of grains.*
CREDIT: OSCAR H. WILL III

Chickens in the Yard

Whether you live in town or out in the country, keeping a small flock of chickens in the backyard is not only fun, it's rewarding in a number of ways. As omnivores, chickens will gleefully seek out and devour all manner of insect, bug, grub, larva, worm, mouse, etc. They will also mow your lawn — to an extent, anyway. Chickens relish fresh greens, including grasses and forbs. When they are confined in relatively small areas, they can keep the lawn trimmed (though, when left to their own devices, they have a tendency to overgraze their favorite things, like clover and dandelions, and spend less time on the Kentucky bluegrass). If you enclose your birds in a portable pen, you can move it around the yard in a rotation, and your chickens will do a much more uniform mowing job than when they are completely free-ranging. So, moving them around in a pen can either keep the birds from overgrazing their favorite vegetation or it can encourage them to do just that — to help you prepare a new garden patch. As the chickens graze, they'll fertilize the lawn with some of the finest organic material out there, but they'll do oh-so-much more.

If you are a lawn purist, you might de-thatch your yard every spring. This arduous task involves hard, soil-scratching raking that pulls up the thatch of dead grass that collects just above the soil surface each year. Alternatively, you might rent an expensive gas-guzzling power de-thatcher that will scratch the soil, while bringing all that dead grass to the surface for easier collection with a leaf rake or power vacuum of some kind. In either case, you are expending all kinds of calories to undo something that mowing redoes every year. Plus, de-thatching can make some turf grass crowns more susceptible to various pests. Here's where the chickens come in.

When left to their own devices, hens will scratch the ground looking for worms, grubs, and other likely food sources. When given plenty of space, or moved around in portable (and bottomless) pens, this scratching will de-thatch and aerate the lawn while breaking the thatch into smaller, more easily decomposed pieces. The end result is that the chicken de-thatchers will render the thatch gone and promote its decomposition in place. You don't have to collect the debris and send it to the landfill or put it in your compost bin. Plus, you can employ chickens year-round to keep the thatch under control. At the same time, they'll keep the lawn fertilized and help control grubs, bugs, and ticks. Chickens do all this and more for the price of a little bit of feed.

Even if you keep sufficient chickens to handle most of the mowing, you might still choose to mow the front yard more formally. Many people who mow with machines collect their grass clippings in black plastic trash bags which are then dutifully sent to the landfill every week throughout the summer. It's true that some folks add the clippings onto their compost pile, but those piles often turn into stinking anaerobic messes because clippings have a relatively high protein content. But there is completely different way of accomplishing the task of mowing. Instead of using machines, you can take advantage of the fact that chickens like their greens. They are more than willing to help you get rid of your grass clippings. (You'll only want to use chickens to mow if you refrain from applying synthetic fertilizers, herbicides, or pesticides to your lawn. Although some folks say that there's no harm in feeding greens fertilized with synthetic fertilizer, we say, don't do it.)

Feeding grass clippings works best with chickens that can be confined — even temporarily — to a spacious pen that has one side or corner devoted to the compost pile. (If you live in town where you aren't allowed to have a compost pile, call it a "chicken feed pile.") As you collect your grass clippings, simply dump them into the pen. You can alternate dumping sites if the chickens aren't eating, scratching up, and aiding decomposition of the clippings quickly enough. Spread them out more thinly if there's even a hint of anaerobic stink going on deeper in the pile.

You can also feed your flock of clucking composters vegetable and fruit waste from the kitchen and garden. The key here is not to overload the chickens. *They* won't mind, but your neighbors might not like the smell, and the code enforcement officer will likely conclude that those chickens of yours stink, when it is actually the vegetable matter. Either way, at the end of the day — month, more likely — you will wind up with a ton of composted clippings mixed with chicken manure and other good stuff that you can spread on your lawn in lieu of store-bought weed-and-feed that really does nothing but make more work for you. And don't forget, even if you do plan to eat eggs or meat from your chickens, allowing them to help you out in the yard will go a long way toward obliterating their feed bill.

Chickens in the Pasture

Much to-do is made about free-range chickens these days. Most people imagine chickens roaming peacefully on lush pasture. But the term "free-range" can mean anything from no cages (but crowded indoor conditions), to free access to a concrete yard, to being raised completely outdoors with little more than a mobile shelter to keep them warm and dry during inclement weather. Especially in the case where the birds are "free" to range inside a chicken production barn, the label is just a marketing scam.

While the completely free-range model is attractive, it is often not practical. The birds might not agree that a barbed wire fence or hedgerow is their boundary, and they are often highly prone to predation. A more practical and humane choice is a free-range model that incorporates some kind of mobile enclosure, complete with predator-proof shelter. You might be wondering: why raise chickens on the pasture at all, except to lower your feed bill in the production of eggs or meat?

When you run cattle through a pasture in a controlled manner, they don't eat everything, and they don't necessarily eat it down evenly. And, while the action of their hooves can help decompose thatch, their manure patties become fly-breeding weed patches if left to rot on their own. Those weeds and the patties represent a concentration of fertilizer that would be better utilized if it were spread more evenly over the pasture. We already know that chickens like to scratch the ground — that's great for the pasture in general. And the chickens will also eat some of the plant material left behind by the grazing cattle. But, even more useful is the way they obliterate manure patties in search of seeds, germinating plants, grubs, fly larvae, and flies. And they distribute all that material in the form of the fertilizer they drop throughout the pasture. No doubt about it, a pasture that welcomes chickens for a fixed interval after the cattle (and/or sheep, etc.) is healthier, more diverse, and freer of flies, grasshoppers, ticks, and other invertebrate pests.

Fig. 1.3: *An easy-to-build chicken tractor can contain your birds when using them to cultivate the vegetable garden and protects them from predators while foraging on range. Raise your dogs among the flock and they'll soon learn not to attack.* Credit: Oscar H. Will III

Managing chickens on pasture generally involves movement of a portable laying or broiler house to fresh pasture every day or two. If the birds are tightly bonded with their structure, they may only roam a hundred yards away from it. If you have light predator pressure, managing this way can work quite well. If you have more predator difficulty and want to limit the size of the chickens' territory, you will want to enclose the birds in large "chicken tractor" pens that have an integral shelter of some sort (and include nest boxes, if you're working with a laying flock). Typically, these pens are moved once or twice per day; larger operators employ a flock of them on pasture. A second alternative is to surround your mobile range shelter with sufficient portable electric net fencing to give the birds the range space they need while keeping ground predators out. This method will not deter any but the most timid of hawks; however, if you can house a chicken-friendly dog along with your birds, you will go a long way toward solving a hawk problem while using the relatively large area, open-top electric netting system.

Chickens in the Garden

As you might already imagine, due to their natural scratching and bug-eating tendencies, chickens have a place in the garden. That's true, but, because chickens also love to eat tender young vegetation (fruits like tomatoes and grains like wheat), their services as gardeners need to be employed a bit more carefully in some cases. Don't let this need for more careful management turn you off, chickens can do much of the legwork involved in building humus-rich soils, keeping pests at bay, composting garden mulch and waste, and post-harvest gleaning.

Consider a typical four-season garden scenario. During winter, you can use your garden as a temporary chicken run — if you have a good enclosure or easily handled portable enclosure that can be moved around in the snow. Winter is a good time to spread hay or straw for the chickens to work into small pieces. And since you must feed your birds through the winter, you'll save yourself some collecting and spreading of manure if you simply let the birds do it for you right there in the garden.

In spring, the chickens will gleefully consume, trample, and generally dispatch any green manures you may have planted in the fall or late winter. They'll continue to work hay and straw down into a friable mulch, and they'll stir the soil surface to aid with seedbed preparations. When you're ready to plant, it's time to pen up the hens a bit more tightly, though. Many folks build chicken tractors that are sized to travel down the garden paths — so their garden hens can keep the paths weed free and well mulched. Others build tractors the same size as the garden beds (raised or otherwise) and move them onto the beds as crops are harvested. Choosing these options will make your chicken-tractor rotations more rational and orderly — but if you're not into orderly, by all means make your tractor the way desire or necessity dictates, and just have fun with it.

Let's say you have one tractor that's sized for paths and one sized for beds.

Fig. 1.4: *Surrounding buckwheat with electric net fencing keeps Freedom Ranger broilers safely contained, while they convert the cover crop into valuable meat and fertilizer.* Credit: Nathan Winters

You could move the path-maintaining tractor around the garden (or into the yard) as required. And you could move the bed-sized tractor from bed to bed, allowing the chickens to prepare the ground for planting by converting hay, straw, grass clippings, etc. into mulch that will later get incorporated into the soil. Later in the season, you can move the bed-size tractor to harvested beds to allow the chickens to glean, clean up any remaining bugs, and help ready the ground for the next crop or cover crop. You can use the chicken tractor to mow down mature cover crops, and so on. The downside with chicken tractors in the garden is that you can't use them to get much help cultivating young crops or controlling bugs in maturing crops.

Some folks use a combination of chicken tractor(s) and chicken moat in their gardens. In theory, the moat model works like this: Create a more-or-less permanent chicken tractor (covered run) all the way around the garden and populate it. The moat should be at least three feet wide, and you can use it as a location for the birds to process compostables as well. In theory, the chicken moat will keep most crawling pests from migrating to the garden because the birds will pick them off as they make their way through the moat. It's just another good way to get useful work from the birds.

Chickens can also be used quite successfully to keep certain crops relatively weed- and bug-free if you let them roam freely in the crop. In these scenarios, you'd typically fence off the crop in question from those that the chickens will damage. For example, you can turn your hens into corn, okra, asparagus, sunflowers, potatoes, and other crops once the plants have gained sufficient height that the chickens can't damage the fruit or tender new growth.

Note: Potatoes don't generally fruit above ground, and the birds aren't fond of the leaves, but some food-safety experts caution that digging root crops in close proximity to fresh manure can increase the likelihood of bacterial contamination, some forms of which have been quite deadly in recent years. We can thank industrial agriculture's overuse of antibiotics in feed, the overfeeding of grains to grazing animals, and other practices that all pretty much point to poor sanitation — a lack of animal husbandry, actually — for those superbugs. Frankly, the likelihood of contracting any serious disease from letting your chickens run in the garden is pretty slim.

The bacterial contamination caveat notwithstanding, you can turn your chickens into your corn patch with little worry of making anyone sick. The birds will enjoy the shade and will feast on the young weeds and myriad insects and caterpillars they're likely to encounter. Some of the more aggressive hens will figure out how to fly-walk up the stalks as the ears fill. If you observe this behavior, simply move the chickens elsewhere. At that point, your corn crop is pretty much assured — so long as you have a raccoon-control method in place and aren't inundated with grain-robbing migratory birds.

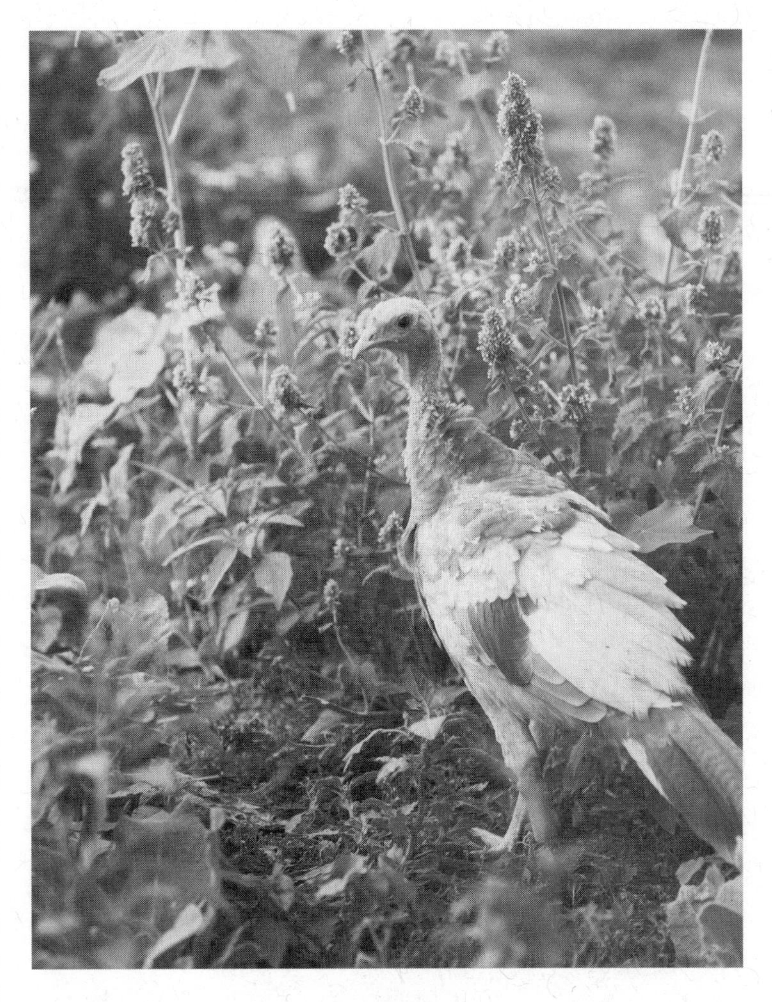

Fig. 1.5: *Heritage Regal Red turkey hen takes a quick survey before ducking into her well-camouflaged nest — she has hatched clutches of up to 12 poults.* Credit: Karen K. Will

Turkeys on Patrol

Even nowadays, the local café's table of truth will offer warnings about how "stupid" turkeys are. "You know what a turkey poult is thinking about from the moment it hatches?" the wise one will ask. "Finding a way to die." "I heard turkeys will drown in a rainstorm by looking up with their mouths open," his buddy will chime in. Exactly how turkeys got such a lame reputation is anyone's guess. Perhaps it's merely a justification for the deplorable conditions most table turkeys are raised under in the age of "animal science." Perhaps it resulted from unsuccessful attempts to raise modern over-bred turkeys outdoors in a self-perpetuating manner (large-breasted commercial turkeys cannot reproduce on their own because their "parts" simply can't mesh). Domesticated turkeys have a very real place on the 21st-century homestead, just as they did those many years ago when all farms were widely and wonderfully diversified. However, as with most of the animals discussed in this book, breeds capable of thriving outdoors make the most sense. You can raise broad-breasted turkeys outdoors, but you will have to learn to inseminate the hens artificially or forever be dependent on others to supply the poults.

Like chickens, turkeys will help with de-thatching in the yard, although they are not quite as efficient at it. They also relish scratching through cow pies, garden debris, and leaves in search of small fruits, seeds, and especially bugs. Turkeys are excellent hunters, and they relish insects like grasshoppers and other pests like ticks. And yes, they will annihilate a rodent nest in search of furry little treats whenever they get the chance. Turkeys are much more territorial than chickens; a flock will tend to gather round curiously whenever you or someone else enters their territory. They will also get after dogs, cats, and other predators until actually threatened by them. Free-range turkeys tend to roam farther than chickens in a given day — in some situations, turkeys head out from their home perching area and may travel a quarter mile or more in search of grasshoppers before heading home for the evening. This can be a good thing or a bad thing depending on where you live and how much daytime predator pressure you have.

Turkeys have a strong desire to perch above ground at night. Despite their size, they usually find some way to fly up to low tree branches, gate tops, front-porch railings, or similar evening roosting spots. If you keep them penned in an area with feed, water, and some nice, covered roosts (about four feet off the ground), they won't perch where you don't want their manure to accumulate. Some folks clip their turkeys' flight feathers to keep them on the ground. Others keep them in large turkey tractors and move them around. Bottom line: turkeys offer real labor around your place in addition to bug control, entertainment, food, and even income.

Avian healthcare folks always caution people to avoid raising turkeys and chickens in close proximity because turkeys suffer from a disease called black head that chickens often carry. You'll definitely want to take this into consideration and avoid mixing your turkeys with your chickens, but thousands of small, integrated outdoor flocks continue to thrive. For best results on the 21st-century homestead, raise heritage turkey breeds such as Midget White, Bourbon Red, Standard Bronze, Black, Royal Palm, and virtually any others listed at the American Livestock Breeds Conservancy website. Because of their somewhat broader view of space, turkeys are not suited to urban, suburban, or other small backyard settings.

What Waterfowl Can Do for You

You don't need to have a pond to enjoy waterfowl on the farm (or even in the backyard), although too many ducks and geese can convert a yard to mud in short order, if they aren't managed carefully. However, ducks and geese can provide all manner of homesteading culinary and craft products, as well as serving more general duty.

If you have room for a small wading pool or plastic concrete-mixing pan, you can easily keep a half dozen ducks or a pair of small geese. Most domestic waterfowl can survive without access to swimming/bathing water. Without it, though, they might not be able to mate because water is an important factor in the equation. And they will most certainly seek out water sources in an effort to live out their genetic destiny. Puddle-starved ducks might take a rain puddle in a low spot in the yard and quickly convert it to a shallow mud hole if that's all they've got. Likewise, geese will muddy up your shallow sheep-watering troughs if they don't have a place to call their own. If you can supply their aquatic needs, keeping ducks and geese is a good modern homestead fit.

Ducks and especially geese will graze for much of their nutrition — at least during the growing season. If you

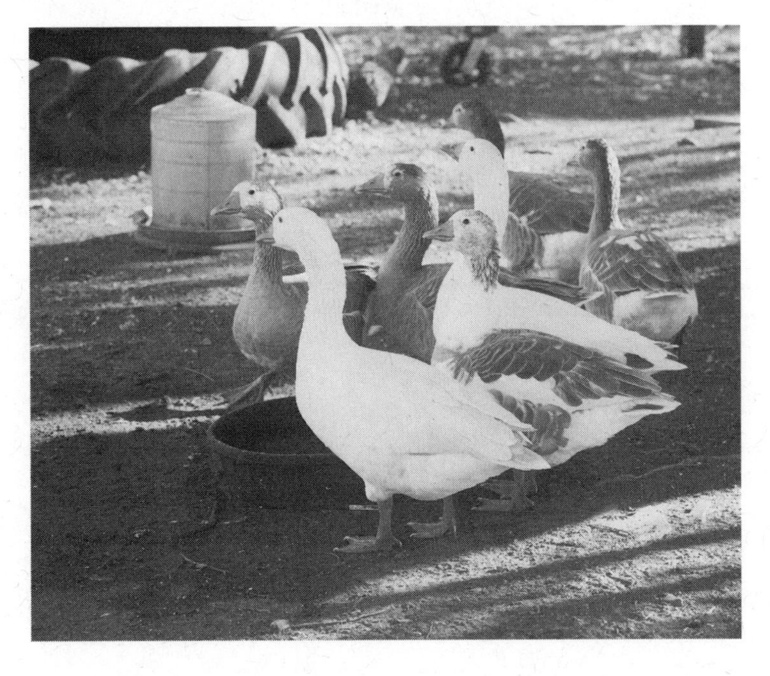

Fig. 1.6: *Geese tend to form tight flocks and can be quite aggressive, especially during breeding season. Watch them carefully if you introduce chickens or other birds in their territory.* Credit: Karen K. Will

Fig. 1.7: *Ducks are even fonder of water than geese and wherever they find water, mud will soon follow.*
CREDIT: KAREN K. WILL

have a small patch of yard that you don't want to mow, a small gaggle of geese will keep it quite trim. The downside is that they will also deposit their nutrient-rich manure as they graze. (So it might be best to let the geese graze where you don't intend to play croquet.) Being grazers, geese are valuable for keeping weeds down among well-established, tall row crops. Some geese have been bred specifically to weed difficult-to-cultivate crops such as cotton (the Cotton Patch breed). Geese are also excellent watch animals and sentinels. They'll approach intruders with necks extended and a deafening babble that's intimidating to the uninitiated human, most dogs and cats, and many other predators. Ducks are more vulnerable to predators, but they have even more to offer you.

Ducks are also good grazers. They so relish fly larvae that they'll muddle through cow pies and other likely grubbing grounds. Some ducks will chase flying insects (near the ground) to good effect, but what they really love is slugs and land snails — it is for this that gardeners the world over quack about ducks. Ducks are also known for their delicious, nutritious, and plentiful eggs. Depending on the breed, a hen duck can rival a hen chicken for egg production. (Bakers prefer duck eggs for certain uses because they add richness and volume.) Ducks are every bit as vulnerable as chickens to aerial and ground predators unless they have access to a large pond.

If you have sufficient facilities to keep their stocking rate low enough that they don't graze patches bare, destroy desirable garden plants, or otherwise get into trouble, waterfowl make a fine, winter-hardy addition to any diversified operation. Ducks are more suitable in urban settings — geese are completely at home on the farm.

Fig. 1.8: *Guinea fowl have a tendency to be a little less cooperative about coming home to roost, but with sufficient imprinting, you will be able to keep them on your place.*

Credit: Karen K. Will

Guineas on Alert

Native to Africa, Guinea fowl have been fancied for sufficient time that domesticated versions come in many different colors and color pattern combinations. Guineas may not be for everyone, mostly because they tend to be quite noisy. This is a drawback in town, but a welcome alarm system in the country. Guineas will announce your coming and going the same way they'll alert you to a coyote that's sneaking toward the chicken yard — with a chorus of cacophony.

Unless you work to tame them, Guineas tend to be more independent, even wild, compared to most chicken breeds. Guineas have been known to roost high in trees, on barn-roof peaks, and even on power lines. So wily is the Guinea that, in some situations, it is capable of keeping its own flock going with minimal human intervention. Obviously, if you are its principal supplier of feed and water, it will remain more domesticated. If you spend quality time with them, Guineas will readily coop up at night, roam the yard during the day, and supply you with meat, eggs, entertainment, and some of the best pest control you can get. Guineas are perhaps best known for eradicating ticks, killing (and eating) snakes, and making life annoying for the local rodent population. These birds are a great fit for the 21st-century homestead. Take care, though, that they don't become *too* independent.

The Mechanics of Chicken Raising

Chicken Tractor

In the same way a tractor systematically plows a field, chicken tractors can be used to cultivate garden patches in a more-or-less controlled fashion.

Housing flocks in tractors is one method used for raising chickens, and the model has evolved to include mobile pasture pens for all manner of fowl, and even some grazing mammals. Chicken tractor designs are as plentiful as there are folks who use them, but most consist of a lightweight metal, wood, or PVC-pipe frame enclosed with mesh wire on at least the sides, ends, and top. Many such tractors have a fully-enclosed shelter with roosts incorporated at one end—some have a nest box in the shelter to facilitate egg collection. In areas with very high predator pressure, chicken tractors might need wire mesh on the bottom also — although this will impede scratching and grazing to some extent.

Basic Chicken Husbandry

Whether your chicken-keeping approach is hands-on or laissez faire, our purpose here is to encourage you to welcome fowl to your farm. We encourage you to try the methods that make sense in your particular setting rather than blindly following some list of rules from some expert. Remember, chickens and humans have been together for thousands of years; for most of those years, the birds needed little in the way of human intervention. Here are the fundamentals:

- Supply a safe and comfortable coop or other living quarters. Minimum space requirements for chickens range from 3–4 square feet per bird for open housing, to 7–10 square feet per bird for confined housing; the size depends on the breed. Go with the smaller size for light breeds such as Leghorns, Ameraucanas, or Hamburgs; use the larger size for heavy breeds such as Jersey Giants, Orpingtons, or Barred Rocks. Don't neglect ventilation in summer and winter, and be sure to supply sufficient perches for all the birds (for details, see section below, *Build a Coop*).
- Supply a clean and continuous supply of water. In freezing temperatures, fill buckets or pans with warm water at least twice a day; a heated fount is even better.
- Supply scratch grains and some relevant feed when the birds don't have access to natural feedstuffs. Broilers should get a broiler ration, egg-laying birds require feed suited to their needs, and chicks need feed that will help them grow without an overload of calcium. If you let hens raise the chicks, you needn't worry as much about supplying any specially formulated chick-growing feed.
- Provide an area for dust bathing. Chickens will create their own dust spas. If you are in an area where lime dust is easy to come by, use that. A lime dust bath supplies free-choice calcium to the birds, as well as help with controlling external parasites.

- If true free-range isn't an option for your flock, offer several outdoor exercise areas and rotate them into use — mostly for sanitation, but also for nutrition, if the areas are vegetated.
- Provide well-ventilated shade in the hot summer and well-ventilated housing in the winter. Heating a coop in winter is not generally recommended unless you live in the most extreme areas and have very few birds in a large, uninsulated space. Never close a winter coop up tight — ventilation is more important than temperature.
- Even on pasture, offer your chickens feed suited to their age and type.
- Keep the coop dry and clean. Any ammonia smell means trouble. For deep-bedding coop management, keep adding dry litter and allow the chickens access to the area beneath the roosts to help work the manure into the bedding.
- You want to remove manure buildup before you detect even a hint of ammonia. Deep-bedded coops should probably be cleaned once a year, if for no other reason than to collect the brown gold for your gardens.
- Spend time with your chickens. Watch them peck, listen to them talk; get to know what's normal and what's not. Most of all, enjoy your birds.

Build a Coop

Before you place that call to your favorite hatchery, you need to think about housing — it doesn't take long for chicks to become chickens! With a little planning and some DIY acumen,

Fig. 1.9: *All you really need to build a chicken coop is some ingenuity and a few common farmstead materials.*
Credit: Karen K. Will

you can set yourself up with a low-maintenance coop using materials you just might have lying around. But whether you are building with recycled or new materials, you need a plan. You can get one from an "animal housing"-type book or the Internet, or you can draw one up yourself. Don't underestimate the importance of this step. Believe us, you'll really appreciate it once you've begun building, and you'll save plenty of money to boot.

All coops need the basics: protection from predators and extreme weather, nest boxes, and roosts. Build your coop to withstand weather at its worst and to protect the birds from ground and flying predators. Build the coop big enough for your flock, but small enough to be maneuvered into its resting spot (or build it on site). Build the coop with ease of cleaning in mind and with easily accessible nest boxes to gather eggs. If possible, place the coop in a wide-open space — not under trees, near fence posts (where birds of prey might perch before nabbing a chicken), or adjacent to heavy brush or woodlands — all predator habitats.

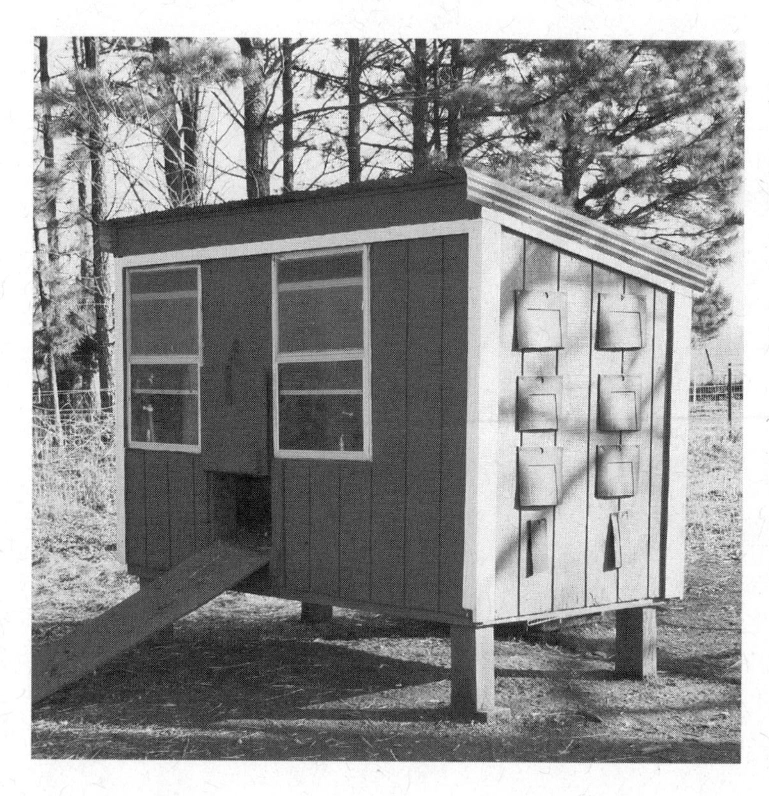

Fig. 1.10: *The authors' layer coop was entirely constructed with recycled materials — nothing like a fresh coat of paint to spruce things up.*
CREDIT: KAREN K. WILL

For construction materials, you can use whatever you have lying around. Consider a combination of materials. You could use scrap lumber or plywood, cordwood, native wood from downed trees around the farm (such as Osage orange), cedar deck boards, hay or straw bales, wood or vinyl siding, PVC pipes (for framing the structure), vinyl or linoleum flooring (for easy-clean coop floors), asphalt or cedar roof shingles (leftover or purchased as broken bundles), corrugated metal roofing or siding, Plexiglas or translucent plastic or fiberglass sheeting (used for windows, walls, or skylight), screening, cattle or hog panels, welded-wire mesh, concrete blocks, or tarpaulins.

For the door, you can get really creative. Maybe you can finally put to use those old double-hung windows you found in the barn rafters (or for free, on Craigslist). Or what about that old door you saved, and the latches from that vintage refrigerator you couldn't bear to send to the dump? Even old, big bushel boxes and wood pallets (use for slat flooring, roosts or steps) are useful. Recycling is the name of the game on the homestead, so don't overlook anything. Consider *everything* in your search.

Hens prefer a dark, protected place to lay their eggs. Rather than a large "community" box, install multiple, small boxes for your laying hens, which turn out cleaner eggs and discourage egg eating. You can make nest boxes, buy them from a farm supply store, or simply use old crates, five-gallon pails, or kitty litter buckets or boxes. Size the boxes according to the breed. (For typical Leghorns, a 12″w x 12″d x 9″h box will do.) A low ceiling provides the cozy cave that hens desire, and it prevents the birds from standing in the nest. (If they can stand up in their box, they'll foul it and then scratch through the bedding.) Bed the boxes with about two inches of hay, straw, or shavings to prevent eggs from cracking when laid. A good coop design will provide exterior access to the nest boxes to make egg collection easy.

Chickens need a high place to sleep at night, so outfit your coop with a roost, or series of roosts. Think "natural" when it comes to roosting material — rounded edges are easier for birds to grip. How about a sturdy tree branch, an old ladder, a broom handle, or a wood dowel? File smooth any rough edges and round off square edges. Allow one linear foot of roost space per bird. Install roosts starting at two feet off the floor with at least a 12-inch vertical and horizontal separation. Because chicken droppings are subject to the law of gravity, don't put roosts directly over each other or directly over the feed area.

Seal any of the coop's unintended openings with ¼-inch mesh hardware cloth; it allows ventilation while excluding even the smallest mice.

Build the coop to sit off the ground by at least one foot; this way, you create a gap large enough that rodents, skunks, raccoons, or possums won't feel safe spending time below your coop. Rest the coop on concrete or pressure-treated wood blocks.

Mesh — woven-wire or electric netting, five feet high — makes the best poultry fence. Bury wire-mesh fencing 6 to 12 inches below the ground (use 6-foot fencing), bent outward to thwart digging predators. Avoid standard chicken wire — it's flimsy and won't hold up against predators or weather.

Raising Broilers on Pasture

Raising broiler chickens is quite a different proposition from raising layers. The main difference is that broilers are temporary residents, whereas layers are more or less permanent.

Modern broiler chickens such as Cornish Rock crosses and Freedom Rangers are full-grown and ready for processing at about 10 weeks of age. You can raise a small batch of broilers each spring or fall on a piece of ground that needs fertilizing (such as a corn patch); this way, the soil gets fed, and eventually, so do you.

Order chicks from a reputable hatchery. Your broiler chicks will be shipped when they are one day old. Don't fret if there are a few casualties in the box; this is normal.

Set up the hatchlings indoors in a brooder. (This will be covered in more detail just below, in the section *Brooding Chicks*.) For the first three weeks, in feeders that are low to the ground, feed broilers a starter ration that is 20 to 24 percent protein. Food and water should be checked twice a day and replenished as necessary.

At about three weeks of age, when the birds have fully developed feathers and have lost their yellow chick fuzz, they can be moved outdoors for finishing. Be sure to provide shelter and protection from predators. (We give you some details on how to do this in the section on page 9, *Chickens in the Pasture*.)

Prepare for Predators

Unless you build the poultry version of Fort Knox, or imprison your birds in a bunker with a cement foundation, you will lose some to predation on occasion. The best strategy for stopping thieves is vigilance. The list of poultry predators is long, but they all leave a calling card.

- **Birds of Prey** (hawks, eagles, owls)
  Most birds of prey have the ability to carry off a small bird (young or bantam). All you'll find is a few feathers. Owls and hawks will enter barns or coops through small openings or fly through windows; they've been known to sidle up on the roost next to sleeping chickens.

  If you find a bird with its head and neck missing, the killer may be an owl. If you find just feathers scattered near a fence post, the thief could have been any flying predator who perched on the post just after the attack.
- **Coyotes**
  Coyotes usually hunt just before dawn and just after dusk, though late in the fall they also hunt during the day. They are sometimes seen trying to break into chicken pens. Coyotes generally take a bird and leave. If they encounter small chicks, though, they will swallow as many as they can until they get nervous and bolt. On pasture, they will take whole, mature birds.
- **Foxes**
  Foxes will typically drag off a whole, free-ranging bird, but rarely enter the chicken coop.

- **Weasels** (includes ferrets, fishers, mink, martens)
  Weasels like to kill for fun, sometimes hunting as a family and doing tremendous damage in a short time. They can squeeze into housing through very small holes and will sometimes reach inside a pen and rip off the head and neck of a chicken. If you find carcasses piled up neatly or birds with their intestines pulled out, you were probably visited by a weasel. Acrid smell in the air? Yes, it was definitely them.
- **Bobcats**
  Bobcats will bite off the head and leave puncture marks on the neck, back, and sides.
- **Wolves**
  Similarly to coyotes, wolves will take whole birds on pasture.
- **Snakes**
  Snakes will devour chicks and eggs without leaving a trace.

Commonsense Tactics

Keep your farmyard — especially the areas around your chicken pen — clean and free of debris where predators hide and rats build nests. Eliminate wood piles, construction materials, hay mounds, and the like.

- **Feed**
  Store all animal feed and birdseed in rodent-proof metal containers. Rats and mice will chew through the toughest plastic, and they'll multiply and nest quickly once they have a regular food source. Don't leave dog or cat food out at night for your pets — it will draw nocturnal predators like skunks, raccoons, and opossums. Look around ... pick up fallen fruit from trees and move bird feeders away from the chicken area — there will always be spilled seed under them to draw unwanted visitors.
- **Fencing**
  The right fence will keep predators out and chickens in. The best chicken fencing is always mesh; it can be woven wire or electric netting, preferably five feet high. Bury non-electric fencing 6 to 12 inches below the ground, bent outward to deter digging predators. If your poultry yard is small, install a "roof" of wire mesh to foil flying predators. If it's too big for that, crisscross wires over the top. From the wires, hang old CDs, which will twirl and twinkle and thus scare off any flying predators.
- **Housing**
  Seal any unintended openings with ½-inch or ¼-inch galvanized mesh hardware cloth. This will keep even the smallest mice out, yet still allow ventilation inside the coop.

  If your coop sits on the ground, bury hardware cloth six inches below its floor. This will discourage and prevent digging predators, like weasels and rats, from burrowing into

Roosters

The question we get over and over is: "Do I need a rooster if I just want eggs?" The answer is yes, most definitely.

Roosters are like good men; they protect their women (the hens) and are first on the scene of any disturbance. A rooster will confront, fight, and most often lose his life when a predator attacks, preserving the lives of your laying hens. Roosters vocalize perceived threats, alerting the flock to imminent danger, such as a flying hawk overhead.

Besides serving as flock protectors, roosters do a good job of marshaling their hens — calling them over to a new and exciting food source, and keeping them together and attentive — thereby reducing harm that may come to them.

Fig. 1.11: *An experienced older rooster will encourage his hens to scratch for food, while keeping a watchful eye out for danger.*
CREDIT: KAREN K. WILL

the coop. If you use mobile housing like a chicken tractor or ark, be sure to move it diligently, at least every few days, because this will confuse predators.

A bright security light or motion-activated light aimed at a coop can deter predators for a while, but they get accustomed to such devices, so they'll be ineffective over time.

- **Trapping**
  Many predators are fairly easy to catch by baiting a live trap with fresh meat or cat food. However, once caught, relocating skunks, raccoons, weasels, and opossums to some far-off location is wasted effort. Besides simply making the pest someone else's problem, most of these predators are family units, so you'll need to catch them all to make a dent. They're also territorial, so they'll travel with purpose to find their way back home.

Chicken keeping is a popular and rewarding pastime ... eggs, meat, fertilizer ... but you'll need to continually observe and strategize in order to keep one step ahead of the myriad predators.

Brooding Chicks

Raising chicks from scratch — or nearly scratch — isn't all that difficult. Folks have been doing it for thousands of years, and most pulled it off without all the technology we have today.

Whether you purchase day-old chicks from a hatchery or incubate fertile eggs instead, there's little reason to believe that your efforts won't result in one fine flock.

Keep the Heat On

No matter what you start with, the first order of business is to keep things warm. Once you've obtained your 1- to 2-day-old chicks, you need to introduce them to their brooder as soon as you can. The brooder can be as simple as a circular pen constructed with corrugated cardboard for the sides and an infrared heat lamp or two suspended above it. (If your design is rectangular, insert curved cardboard pieces in the corners to keep excited birds from piling up and suffocating.) Cover the brooder's floor with an inch or two of clean wood shavings, ground corncobs, chopped hay or straw, or other absorbent material. As you place the chicks into their new digs, dip their beaks in the water trough (a shallow container like a mayonnaise jar lid will work in a pinch) to introduce them to it, and let them go. If all the chicks converge beneath the light(s) in a tight huddle, they're cold. If they form a ring around the light, it's too warm, and you should raise the light some. Adjust the lamp height up or down until the chicks are more-or-less evenly dispersed below it — happy chicks will peep with a murmur that is mesmerizing. They will come and go, venturing out of the heat to drink or peck in the food trough.

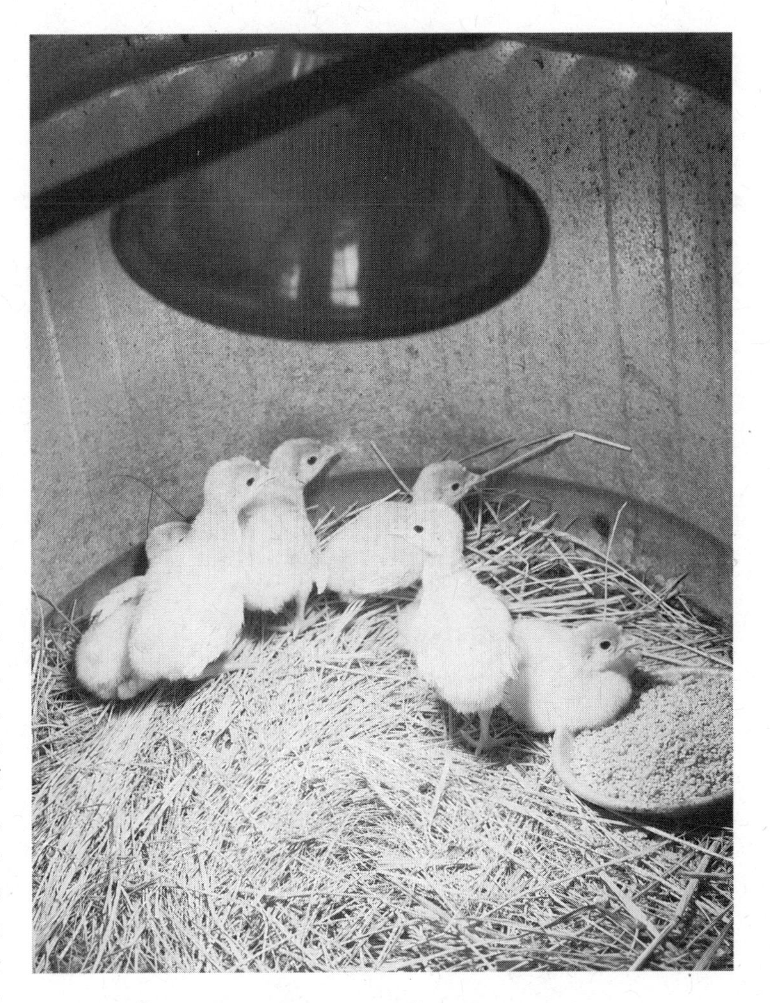

Fig. 1.12: *A chick brooder can be as simple as a heat lamp and a small stock tank or box. These Midget White turkey poults were ready for their outdoor pen at about 10 weeks of age.* Credit: Karen K. Will

Folks have successfully brooded chicks in cardboard boxes, plastic tubs, metal or plastic stock tanks (water troughs), old chest freezers with their lids removed, etc. Take a look around and don't hesitate to put some old container to good, new use — just be mindful of heat and fire hazards. If you

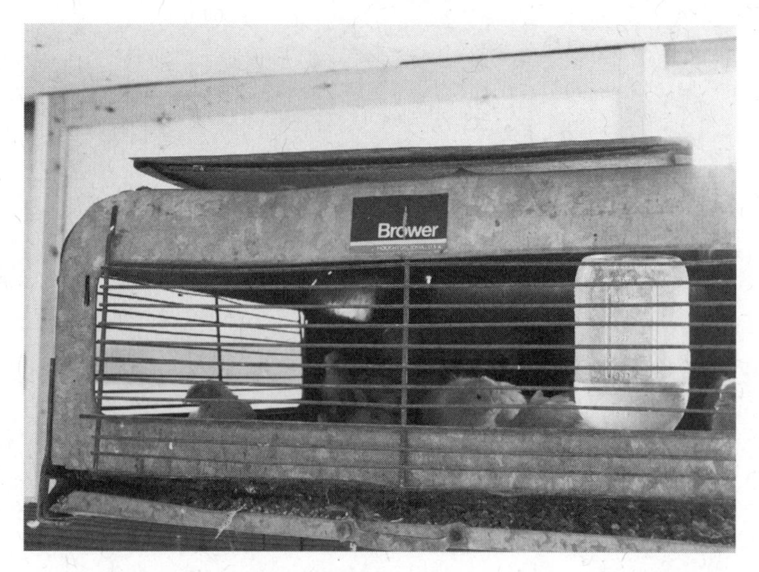

Fig. 1.13: *Commercial chick brooders come in many different sizes and shapes. If you find one at a farm sale, by all means snap it up, assuming it hasn't corroded half way to oblivion.*
CREDIT: KAREN K. WILL

brood in a cardboard box, you will want to be extra certain that the heat lamp cannot fall or in any other way come in contact with the cardboard. Likewise, you will want to be sure that a lamp never comes in contact with bedding, no matter what kind of brooder you use.

Stoke the Furnace

In addition to warmth, chicks require a ready supply of clean water and free-choice food, at least initially. Choose a chick starter ration from your local farm store. (Ask for non-medicated, or you will likely get feed with medication in it). Starter feed should have a protein content in the vicinity of 20 to 24 percent. At about five weeks of age, you can switch your birds over to a growing ration that has a protein content around 15 percent. If you are mixing the ration yourself, be sure to include grit for their developing gizzards and hold off on any significant calcium supplement until they are almost ready to begin laying (around five months of age). Once the birds are fully feathered and it is mild outdoors, you can move them to their coop. If nighttime temperatures dip into the 40s, you might offer a heat lamp in their perch area if they seem uncomfortable.

Keep It Clean

Cleanliness is key to successfully rearing a brood of chicks. Avoid letting their feed get wet, lest it grow mold. Avoid the temptation to fill their waterer without first washing it out — parasites and disease can spread rapidly through contaminated water. Remove and replace wet bedding in the brooder — add more as manure builds up. And finally, don't forget to compost that good stuff when you move the chicks out of the brooder — your garden will thank you.

Have Fun

Don't forget to enjoy your brooding chicks. They will offer you almost as much entertainment as will the adult flock. Some chicks are precocious; some are shy. Some chicks are bullies; some are natural-born leaders. Sometimes the males will engage in mock sparring. Most of the time they will just interact with one another in an interesting and engaging social way that's quite relaxing to watch.

Hatching a Flock

Which came first, the chicken or the egg? For most backyard poultry enthusiasts, the chicken came first — well, more correctly the day-old chicks first arrived in the mail. But that's not the only way to create your first flock or maintain your existing one. For folks who are uncertain about receiving live animals through the mail, or simply cannot handle the minimum number of day-old chicks that most hatcheries require you to order, incubating fertile eggs is an attractive alternative. (You must, of course, have a rooster in your flock if you want fertile eggs.) Incubating eggs is a great way to increase flock size or to provide replacements for birds that have been culled or killed. Hatching fertile eggs need not be difficult, but your success rate can be increased by following a few guiding principles.

Environmental Needs

Chicken eggs need a fairly specific environment to develop properly and hatch successfully. The most important parameter is temperature — chicken eggs should be incubated at a temperature between 99 and 102 degrees F (99.5 is often considered to be ideal) and 50 to 65 percent relative humidity (55 percent is often considered to be ideal).

To facilitate proper aeration and gas exchange between the embryo inside the egg and the outside world, the eggs must not be held in a tightly sealed container.

Chicken eggs typically hatch after 21 days of incubation. Consider that number to be a guideline — not an absolute. During the final three days of incubation, the eggs should ideally be located in a slightly cooler (98.5 degrees) and more humid (65 percent relative humidity or greater) environment to facilitate successful hatching. Lowering the temperature helps account for the extra heat that the larger embryos produce as their metabolism increases, and the increased humidity helps keep them from getting stuck to the membrane that's located just inside the eggshell as they break out of the shell.

Mindful Manipulation

Just as temperature and humidity are important to maximizing the hatch, eggs need to be moved around on a regular basis for best results. Changing an egg's attitude helps exercise the embryo and prevent it from sticking to the shell. In general, eggs should be incubated with their pointed ends down (air cell up) — but it is also important to turn or tip the eggs back and forth at least twice a day — the more often they are turned, the better. During the hatching phase, it's best to lay the eggs on their sides.

Natural Incubation

The easiest way to incubate and hatch fertile chicken eggs is to have a broody hen do all the work for you. What's a broody hen, you wonder? It's a hen that has undergone progesterone-induced changes that make her want to sit on eggs to hatch them and brood the resulting chicks. A broody hen will take care of ventilating and warming the eggs and will handle all of the turning and chick-rearing duties as well. If you already have a laying flock and one of your hens becomes broody, she can incubate her own eggs or you can place fertile eggs obtained elsewhere in her nest and she will do her best to hatch them. Many folks try to prevent their laying hens from going broody because they no longer lay eggs. But if you want to hatch a few chicks, a broody hen can be a godsend.

Artificial Incubation

When most folks think of incubating eggs, their minds turn to any manner of electro-mechanical devices that provide the right temperature and humidity. There are two principal categories of incubators suitable for the home flock: still-air and forced-air. Still-air incubators lack mechanical air circulation. Forced-air incubators use a fan to circulate internal air. Both types of incubators may be equipped with automatic or manual egg turners, and both offer some means for managing relative humidity. Incubator capacity and prices vary widely, so it's wise to consider how many eggs you are likely to hatch in a year before you take the financial plunge. If you choose a small incubator without an egg turner, remember that it is up to you to turn the eggs at least twice a day.

Experts recommend that you set the temperature of your still-air incubator to 101 to 102 degrees F to avoid the formation of cold spots. Forced-air incubators should be set at the specific temperature you desire because the moving air creates a more uniform temperature environment. All incubators should be turned on, adjusted, and monitored for at least a day before you set the eggs. Check the temperature with a thermometer that you know to be accurate — a degree or two one way or the other can make or break the hatch. In a still-air incubator, the thermometer should be placed at about the height where the top of the eggs will be.

Serious poultry enthusiasts may have one incubator for incubating and a second (sometimes called a hatcher) for hatching. This allows you to set eggs at virtually any time (mark them carefully!) because the incubator's environmental parameters won't need to be reset for hatching after 18 days.

Using a hatcher also helps keep the incubator clean.

If you plan to hatch eggs on large scale, you might want to consider a forced-air, cabinet-type incubator. These devices offer a great deal of capacity and flexibility, but they cost from several hundred to thousands of dollars.

Most home incubators are designed to operate effectively at ambient temperatures from the 60s to the 80s. With any hatching project, be sure to locate the incubator in a spot that will be undisturbed and out of direct sunlight for several weeks at a time.

Cleanliness and Storage

If you purchase fertile eggs from a hatchery, more than likely they will have been sanitized. However, if you are going to incubate eggs produced by your flock, or you obtain eggs from a source that does not sanitize them, you can avoid potential health and viability problems with a sanitizing rinse. Using a capful of bleach to a gallon of water that's warmed to about 110 degrees (you can substitute liquid dish soap or the recommended dilution of Tek-Trol), immerse each egg for a few seconds and air dry.

Incubators and hatchers should be cleaned out after every hatch and sanitized after every third hatch, at the very least. Dust or vacuum the interior and wipe all surfaces and trays with a dilute bleach solution (up to a quarter cup per gallon) or other sanitizer that won't leave a residue or emit vapors that could poison a future batch of eggs.

Fertile chicken eggs can be stored up to 10 days (before incubating) with little loss in hatchability — as long as you keep them out of the refrigerator. The ideal storage conditions are 55 to 60 degrees F and 70 to 75 percent relative humidity. Store the eggs in trays, bowls or clean egg cartons with sufficient space to allow air to circulate. Some experts recommend turning the eggs in storage, too. This process can be easily accomplished by tilting the entire egg tray (or whatever container you use) from side to side.

Preview the Progress

Candling is a process that allows you to determine whether your eggs are developing correctly or not after seven to ten days of incubation. Candling is a great way to monitor the progress of your eggs, but it takes a bit of practice to get right.

Although you can purchase devices designed specifically for candling eggs, all you really need is a bright (preferably LED) white-light flashlight and a dark place. Ideally, the end of the egg should "seal" against the light — if your flashlight lens prevents this from occurring, you can make an adapter tube out of cardboard, or cut a hole in the lid of a cardboard box sufficiently small that it will cradle one end of the egg and hold it upright. A lamb nipple with the end cut off and pulled over a good pen light also works well. In any case, illuminate the egg (from below, in the box setup) and look for a web-like network of blood vessels surrounding what is obviously a chicken embryo (by seven days you may notice embryo movement).

Clear space and a yolk, or a ring of blood (just a single ring, not the network of vessels you want to see) indicate that the egg was not fertilized or that it died during the early stages of development. Note: It's not unusual to lose up to 50 percent of the eggs you initially set, depending on the quality of the eggs, the incubator model you use, and your diligence and skill. Eggs that aren't developing properly should be discarded because there's a higher risk that they could explode in the incubator — and that's a mess that no one wants to contend with.

Once the chicks hatch, you can leave them in the incubator or hatcher for a day or so before moving them to the brooder. Newly hatched chicks obtain sufficient energy from residual yolk that really all they need for the first couple of days of life is a warm environment — so there's no need to rush them to the brooder after hatching.

Fig. 1.14: *Portable wire pens, or chicken tractors, make an excellent and safe environment to allow adolescent chickens to develop into adults — be sure to move the pen regularly.* CREDIT: KAREN K. WILL

Chapter 2

Plowing with Pigs

One of the most energy-intensive aspects of converting a piece of sod-covered ground into a productive garden or grain field is getting rid of what you don't want. Actively growing grasses and forbs need to be removed, and you have to find a way to deal with the weed seed bank you are about to open for business. Whether you strip that sod by hand, turn it with an animal- or tractor-pulled land plow, or employ some other expensive, fuel-guzzling machine to get it done, the initial breaking of the land takes a lot of energy. While it's a pretty straightforward task to turn a 20- by 20-foot patch with a spade, turning something close to an acre by hand — and *successfully* killing the unwanted plants — is not a job for the faint of heart.

The race begins the moment the land is laid bare. Even in soil that seems relatively weed free, there is a seed bank — a store of seeds lying dormant in the soil, waiting for an opportunity to take off. When you turn the growing medium, you'll inevitably bring some undesirable seed from this seed bank to the surface — where it will germinate and compete with your garden plants or crops. Over the centuries, humans have invented all kinds of tools to help give crops and garden plants an edge over the weeds. From inexpensive hand hoes and wheel hoes to more expensive rotary tillers to huge and costly tractor-mounted cultivators, beating back the weeds has been on homesteaders' minds from the beginning. Modern agriculturists would have us buy machines and apply expensive mixtures of highly poisonous herbicides to our gardens and crops to keep the weeds down today. Yikes! What can a 21st-century homesteader do?

Simply prepare your ground using pigs.

Old Pigs, New Tricks

Until the relatively recent industrialization of agriculture, pigs were a welcome addition to virtually every farm and smallholding across the country. Back in the day, pigs were more important

Heritage Hogs

It used to be that every farm and many a backyard in town had at least a couple of hogs to put food on the table, if not to add profit to the bottom line. These pigs were raised for their lard (once much more valuable than their flesh) and/or meat, and they were generally adapted to the general locale where they were raised. With the advent of "the other white meat," many of these so-called heritage breed pigs were lost because there simply was no longer demand for an animal whose carcass and behavior couldn't conform to confinement production models. Although today, there are many fewer heritage breeds from which to choose, you can still find pigs that will thrive outdoors and that will put their good snouts to work rooting up the ground, destroying weeds and perennial roots and grubs — all while fertilizing ahead of your crop. Check the American Livestock Breeds Conservancy website (www.albc-usa.org) for likely sources of heritage breeds, or find a local farmer who still raises old-line hogs (such as Duroc) or heritage breeds (such as Tamworth) outdoors.

Fig. 2.1: *Old-line hogs like these Gloucestershire Old Spot and Old Spot crosses were once raised almost exclusively on the rich mast produced by hardwood forests. Acorns and other nuts are still an important part of this herd's nutrition.* CREDIT: KAREN K. WILL

economically for the fat they produced than just about anything else, but they also served up protein (protein that in no way resembled the modern industrial "other white meat"). However, early homestead hogs were much more versatile than we give them credit for today, and the variety in old-time oinkers was vast.

In the latter half of the 20th-century, as chemical engineers searched for ways to make weeding easier (i.e., via poisonous cocktails), animal scientists were breeding leaner, longer, and faster-gaining hogs that would suit the pork industry's move into mass-production agribusiness. Today's industrial pigs are raised entirely in confinement, on concrete slats or floors, and with prescribed rations. They live their lives in a forced and formulaic fashion with a sole purpose: to supply bacon, chops, and hams that fit neatly into boxes sized for efficient shipping around the country. These modern specimens have been so highly bred that the pigs of old appear odd by comparison.

Crowded, confined conditions and slotted concrete floors are hardly reminiscent of the oak forests and legume-rich pastures where pigs once hogged. Humane animal husbandry issues aside, the pigs developed by modern breeding programs wouldn't even be able to take care of themselves in the great outdoors — where they had been taking care of themselves for centuries. The modern breeding program has been so effective and "animal science" indoctrination so complete, that most of today's farmers can't remember the type of pig that offered anything to an agricultural operation other than bacon, chops, and a couple of hams.

If we could visit homesteads and farms of yore, we'd see a much different pig from the one animal engineers offer us today. Luckily, a few of those breeds are still with us. These "heritage breed" hogs are capable of thriving outdoors and are among the most likely to partner with you on the labor required around your place.

The pigs our ancestors raised were versatile beasts that served a multitude of functions around the farm and homestead. Sure, they were a source of sustenance for most people and also supplied a once-thriving, non-petroleum-based lubricant trade with the lard it needed to keep the gears turning and the shipyard ways slick. However, many homestead hogs were expected to dispose of excesses, be it from the supper table, dairy, garden, or field. These old-style hogs were also expected to glean, work soil, kill snakes and other small varmints, harvest mast crops, and perform a host of other tasks for which they were genetically suited.

Although the number of heritage breed and old-line hogs available today is tiny compared to the late 19th- and early–mid 20th-centuries, thanks to the efforts of preservation-minded pig enthusiasts and organizations such as the American Livestock Breeds Conservancy, a few classic breeds are still around, and their numbers are increasing annually. As such, old-style hogs can once again find gainful employment on the homestead in a number of ways — ways that may seem innovative and strange in our techno-crazed 21st-century.

Pigs as Plows

Spend a little time watching pigs outdoors, and you'll soon discover that they interface with the world beneath them through their snouts. Though seemingly tender, the pig's nose, indeed her entire head, is designed to facilitate her most-cherished activity — rooting. Pigs root for food, they root when they are happy, they root when they are excited, and they root out a nice groove in the mud in preparation for a cooling spa treatment.

Fig. 2.2: *Young pigs will eagerly plow up the sod in search of a meal. Most hogs are easy to contain using either electric net fencing or just a few strands of electric wire fence.* CREDIT: NATHAN WINTERS

Experienced pig herders young and old, and, indeed, the sage advice of late 19th- and 20th-century agriculturists, suggest that pig husbandry should include ringing or cutting the pigs' snouts to prevent their rooting because it can damage pastures, lawns, flower gardens, and other areas where soil disturbance is not desired. Thank goodness we've come to our senses with regard to cutting snouts (which was done with special pliers-like tools that cut the cartilage ring that shapes the pig's nose — and done sans anesthesia, of course), and a new, humane ring is available that will interfere with the pig's snout sufficiently to prevent it from rooting. But wait ... what about all those tough homestead tasks that take advantage of the pig's rooting skills? Why alter a pig's nose at all?

It turns out that a patient (and observant, engaged) pig herder can control pasture damage from rooting with well-timed rotations. You have to spend time with your animals, but that's the reason you have them in the first place, right? If you discard preconceived notions and take the time to observe pig behavior, you'll see they are self-contained, land-plowing units that can turn sod and rid the soil of grubs, grass, roots, and rhizomes like no combination of machines and pesticide sprays so far invented. Take it a step further, and watch pigs rooting. Notice that the leading edge of a pig's snout is shaped very much like the chisel plow shank that farmers pull through the ground to shatter hardpan. No, pigs won't plow up the ground in neat rows, but they will plow it up in their own whimsical way, if you give them sufficient time to do it. (A group of five adult pigs easily can plow half an acre in a couple of months.)

Fig. 2.3: *It only takes a few weeks for 5 mature hogs to ready last year's quarter-acre corn and sunflower patch for planting turnips and other fodder crops.*
Credit: Karen K. Will

Using pigs for plowing is as simple as confining your herd (or a small portion of it) in the area where you want the soil turned. You can confine the pigs using welded steel hog panels and fence posts of some sort (see Chapter 4 for ways to find free posts) or, for more flexibility and less overall investment, consider an electric fencing outfit (see below) to keep the animals concentrated where you want them. If the area is a permanent garden or field, you might consider planting and weaving an impenetrable hedge around it using thorny plants such as black locust, Osage orange, or roses, especially if you can get the seeds or seedlings for free (see Chapter 4).

Your pigs need shade and plenty of water no matter where you ask them to plow (don't fret over the land lost to their mud wallow), and you can get them to do an especially thorough job of plowing in an area if you scatter some cracked corn or other treat such as scratch grains. How long to keep your pigs in one area will vary. You need to use your head here — it will depend on how many hogs, what size the hogs are, what breed they are, how thick the initial vegetation is, and the species matrix making up the vegetation. You will know they've completed the job when the soil is free of plant shoots and roots. If the pigs stay longer, they will simply fuss at you more for grain or other feed.

Pigs in the Garden

The pig's place in the garden comes after you've taken everything you want

Fig. 2.4: *Hay also has entertainment value — for your pigs — and they will quickly turn it into mulch and eventually compost.* CREDIT: KAREN K. WILL

from it in the fall. Pigs are excellent gleaners, and they will relish leftover tomatoes, whether overripe or green. They will eat the weeds, tear out the vines, mow down and munch the cornstalks, and root out any remaining root crops. If you used deep mulch on your garden, the pigs will break it up and turn it under. At the same time that they're doing all of this, they'll also be fertilizing the ground. Please note that some food-safety experts suggest that you avoid growing root crops and greens for human consumption on ground where fresh manure has been recently applied — the risk of passing a pathogen appears to be real, even if it is fairly slight. But (as our ancestors did) many folks use pigs to clean up the garden in the fall in preparation for cover crop, then plant food crops as normal in the spring. If it is your flour corn patch one year and your tomato patch the next, any risks associated with fecal pathogens will be minimized because tomatoes and corn are borne well off the ground where the manure was applied.

Making Mulch

Pigs not only effectively plow up the ground and fertilize it as they go, they will also help you mulch. If you provide them with several large round bales of hay, stacks of hay or straw, etc., they'll rapidly tear them apart, break the stems into small pieces and integrate that compost-like mulch into the soil. In cold climes, if you're using pigs to plow in the fall and can overwinter them in the field, you'll need to provide them with plenty of hay or straw as bedding to bury themselves in. The pigs will relish biting, eating, chewing, and rooting in that hay or straw, and, in the process, just a few pigs can easily add several tons of mulch to the ground by spring. And they'll work it into the ground — if it isn't frozen. Sure, you'll need to provide the pigs with a decent winter ration that includes grains, but think of it this way: The money or barter you might spend on winter feed for the pigs will not only keep them healthy and happy, it will also add fertility to the ground where they are living. You'll get that back as you harvest future crops from the plot.

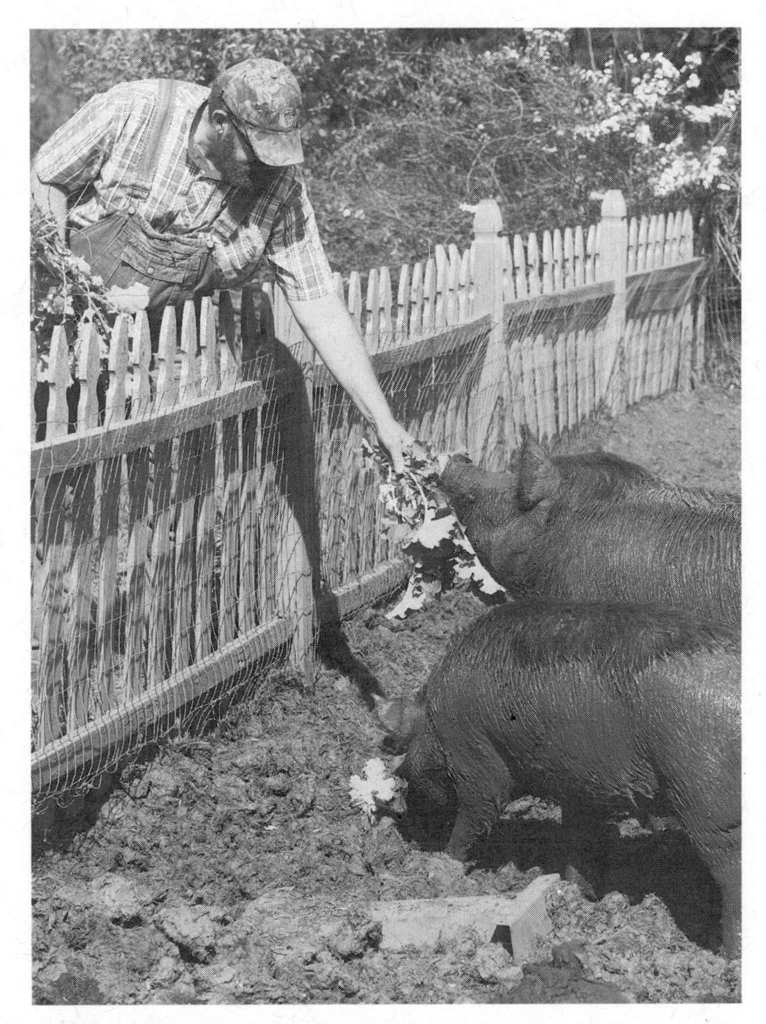

Fig. 2.5: *There's no summertime hog treat so relished as greens freshly pulled from the patch — unless it's being let into the corral to mow down 4-foot-tall lambs-quarters.*
CREDIT: KAREN K. WILL

Ultimate Composters

If you wish to produce relatively large quantities of compost with pigs, it's useful to have a decent supply of both "greens" and "browns." Browns might come in the form of old hay or straw; greens might be manure from your milk cow, fresh grass clippings, or another high-nitrogen material. Ideally, you'll already have your compostables confined in a pig-friendly place, but there's really no harm in carrying the material to the pigs if you can't move the pigs to the piles.

If you winter or simply bed (with hay or straw) ruminants such as cattle or sheep in a barn or small outdoor lot, you should consider trying the following technique that can a save you a huge amount of work:

As you feed the ruminants hay and grain, they'll deposit manure and liquid waste on their bedding. Rather than removing soiled bedding and adding fresh, simply add clean bedding on top of the soiled (this is one of several variations on the deep-bedding model for confined or semi-confined animals).

The mixture of manure, urine, and bedding material may start to compost on its own, but more than likely it will be too cold, and the animals will trample it into a "manure pack" that will be about as tough to break up come spring as a well-made adobe floor. Rather than worry about what to do with the hard pack and/or to speed up the composting, scatter some corn or other grain beneath the fresh bedding every time you add it. Pigs love to root, and although they will root "just because," if they know there's a reward such as a kernel of corn or wheat now and then, look out! The pigs will put their snouts to work and in no time at all, that manure pack and/or soiled bedding material will be broken down into one of the richest soil amendments you'll ever have the privilege of adding to your gardens and fields — and you'll be growing pigs at the same time! And that's not the only strenuous work you can entice pigs to do with yummy treats.

Pigs will also happily help you compost food scraps, even tough-to-compost meat scraps (but avoid feeding pigs pork), although to be truly effective, you'll need to do this in conjunction with the deep-bedding process or with the mulch-making process described above. All you need to do is feed the pigs the scraps, and they'll convert them to pig manure in short order. The pig manure is then mixed in with the other mulch- or compost-making materials and — just like that — last night's turkey carcass is now ready to amend soil. Note that when feeding things like turkey or chicken carcasses, you needn't chop them up first — the pigs will do that for you, and the waste will disappear before your very eyes.

Technical Language

As with any kind of livestock, there is a language that goes along with swine. Although we use the words "pig" and "hog" more or less interchangeably in this book, there is a technical difference that you need to know if you plan to talk swine with experienced swineherds.

- Piglet: term used for baby swine that's rarely used by folks in the industry
- Pig: a young swine — something you might be tempted to call a piglet
- Shoat: an adolescent pig that's been weaned but has not yet hit 120 pounds
- Hog: a maturing swine that has passed the 120-pound mark
- Boar: an intact male
- Barrow: a castrated male
- Gilt: a young female before her first litter
- Sow: a mature female hog after her first litter
- Weanling or Weaner: 8- to 12-week-old pig that's just been removed from its mother
- Feeder Pig: the young animal (generally less than 70 pounds) you might purchase to plow with

Hogging Stumps

If you live on the land and never have to stump some trees, you are a rare bird indeed. The process described here works with any tree. If it is a hardwood tree or a very dense-rooted softwood (larch, hackmatack, etc.) you can make mallets with the wood. Many folks use softwood for heating because that's all they have; and to us, softwood does not imply "not firewood." Our ancestors used pick, axe and shovel, explosives, horse- or oxen-powered winch or lever, and yes, you guessed it — pigs. The pig approach to stump removal is a little haphazard and not so awfully efficient, but if you already have the pigs and your time schedule is flexible, then hogging stumps is just one more way to get something done for little labor or monetary investment.

The first step to hogging stumps is to enclose the protruding tree anchors with a pig-proof fence — hog panels and electric fence work well. Once you have a place to put the pigs, focus your attention on the stump(s) in question and, to the best of your ability, bore holes down between the roots (some folks even auger holes into the stump itself), and bait them with one of the pig's favorite grains, like corn. You might need to leave a few kernels at the surface, but the pigs will dig and chew and dig and push with their powerful heads and snouts to get at all the buried grain — and they will delight in the grubs, worms, and other critters they find as they dig.

It might take a day, it might take a week, it might take a month, but in time, the stump will be destroyed or excavated completely from its hole. If you are lucky and the pigs leave the stump intact, you will have a lovely resource from which you might make mallets to drive your chisels, gouges, or froe, gluts to help split logs, or some really fine, long-burning firewood (see Section 2).

Pigs on Patrol

When raised around other barnyard animals, pigs are pretty peaceful co-existers. Don't worry if your chickens or geese wind up in the pig pasture one day — the pigs aren't likely to attack or try to eat them. If the birds decide to lay eggs in the pig's playpen, well, don't expect any to hatch, or even be around to collect. As omnivores, pigs love eggs almost as much as they love lush green grass, whey, and corn. Pigs will kill and eat unsuspecting snakes that cross their territory, and they're happy to snack on a rodent or ten whenever they get the chance. Since pigs can readily go feral — indeed, most wild pigs in the United States escaped the farm or were released into the wild by unscrupulous hunters — we don't recommend letting any but the most tame of pigs roam your property unchecked. Thus, using pigs for pest patrol is a bit problematic unless you create a moat-like pigpen surrounding your home. Since snakes do more good than harm, and since there are better animals to employ on rodent and other varmint patrol, be content to marvel at the wonder of the hog's versatility when it happens to catch and devour a snake, rat, or mouse, but try not to rely on pigs to keep the varmints in check.

Pig Control

Keeping your pigs under control isn't terribly difficult, especially if you raise them from weanlings and you develop a rapport of trust — this usually involves being a gentle source of treats (grain, turnip greens, fresh clover, apples, eggs, etc.). It's not a great idea to make full-blown pets of your pigs (especially if you intend to sell or slaughter them), but a certain amount of gratification comes with the knowledge that your animals are calm and relatively easy to work. Regardless, you should always keep your wits about you as you work your pigs — 600-pound gentle giants can injure you even if they bear you no malice. Especially with intact boars and sows with litters, pay close attention to how they respond as you approach. Sows sometimes charge if they believe their little ones are in danger, and boars can charge and slash with their tusks if you interrupt a romantic interlude, or even if he thinks you might.

Fig. 2.6: *Mulefoot shoats getting comfortable in their new home. Within 3 months, they had each gained more than 100 pounds and had completely opened up the sod. The hut was entirely constructed from pallets and other recycled materials.* Credit: Oscar H. Will III

The best way to control pigs is with a combination of permanent and portable fencing along with some handheld rectangular devices called sorting panels. Sorting panels are useful in close quarters. You can make them from scrap lumber or metal and use them to turn pigs, block their progress, or protect yourself from intended or unintended (on the pig's part) encounters with an animal's mouth or hooves.

Most pig setups will include a more-or-less permanent central pen area where the group or individuals can be brought in and "trapped." It's convenient to provide food, water, and wallow in this area because the pigs will associate it with good things, and you won't have difficulty convincing them to come on in. This pen is where you might catch piglets for weaning, separate an animal for loading onto a trailer, or just get a closer look. This area might be best enclosed with welded hog panels, although some folks use woven-wire mesh backed with wooden posts and rails.

For pasture or garden enclosures, you needn't bother with such a stout enclosure. For those applications, electric fencing makes a fine choice for containing hogs. You might choose a permanent setup for the pasture perimeter using strong corner posts and three strands of conducting wire spaced at about 6, 12, and 20 inches off the ground. You could add a fourth wire at about 28 inches if you are concerned that the larger animals might try to make a break for it — but after they've been shocked once, pigs rarely go close to the wires.

Temporary electric fencing can be used to create cross fences, separate animal groups, and create alleys leading to an area you want to plow. You can also use it to enclose those spaces. Temporary electric fences usually require a few hand-driven steel t-posts, a handful of insulators, a couple dozen step-in insulated plastic or fiberglass posts, and two conductor wires — one placed at 6 inches and one at 12 inches off the ground.

Both permanent and portable electric fencing can be energized with the same charger. Invest in the best charger you can afford for the total distance of conductor you anticipate using. Bigger will be more flexible, in general, but sometimes you can be better served with a pair of chargers — one that plugs into a 110 outlet and another that gets its juice from the combination of the sun and a battery. The solar/battery-powered unit will make it possible to take the pigs virtually anywhere on the farm without needing to run a quarter mile of extension cord. Of course, if you find a couple of good chargers at a farm sale or on Craigslist, and the price is right, ignore this advice and just do what you need to do to make them work.

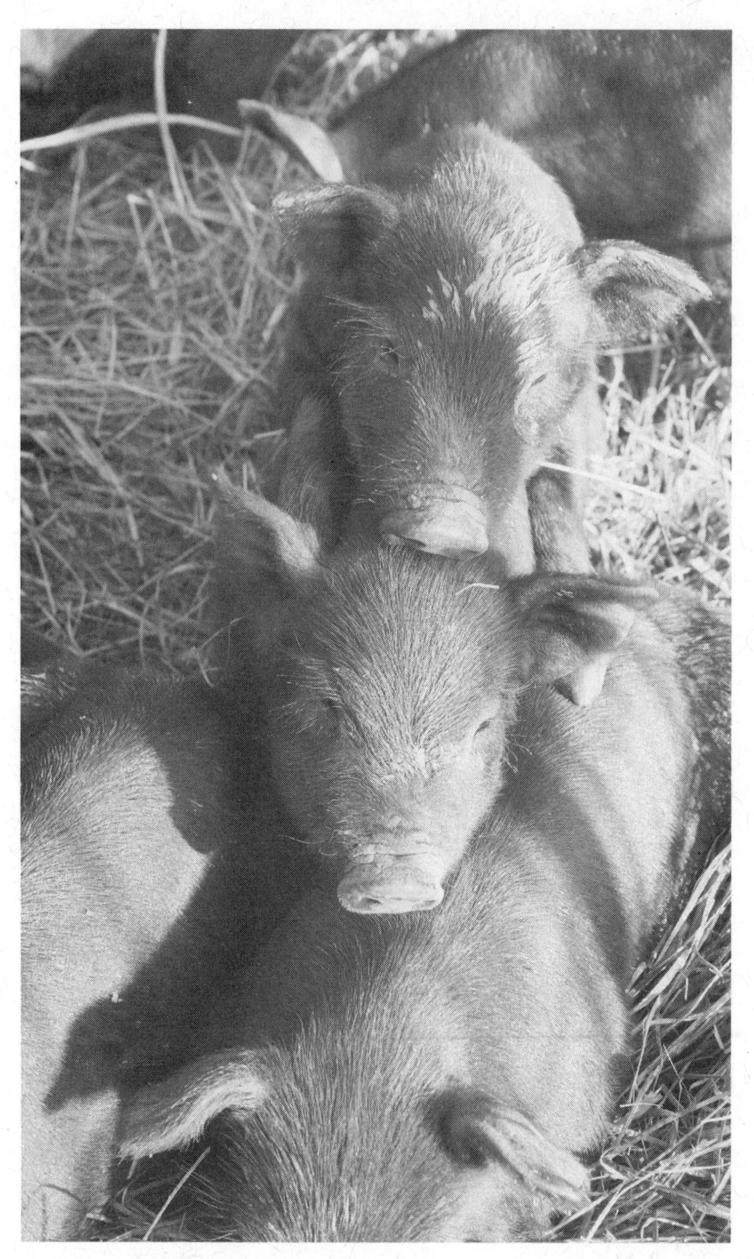

Fig. 2.7: *Piles of baby pigs and endless hours watching their entertaining antics are but one of the joys associated with raising hogs.* CREDIT: KAREN K. WILL

And by all means, read everything you can about raising hogs outdoors and talk to as many folks who do it as you can before you design and install your facilities — or, do as many people do and jump in with both feet, an open mind, and a healthy sense of adventure.

Live Entertainment

Taking the time to enjoy pigs is a large part of raising them in any setting. Heritage breeds, with their quirky behaviors, unconventional builds, and uncanny ability to look after themselves make fine additions to virtually any small agricultural enterprise. No matter your business's model or hobby's goals, these pigs will partner with you as you tend the land, following wherever it leads you next. And even if eating meat isn't in your future, a small herd of pigs adds all kinds of value to your place.

Basic Hog Husbandry

Entire books have been written on the care and nurturing of pigs — some take a more hands-on, meddling approach, while others suggest that the animals don't really need much in the way of intervention. We advise that you try the methods that make sense in your particular setting, rather than be dictated to by a list of rules that one particular expert says is critical. Remember, pigs and humans have been together for thousands of years. For most of those years, the pigs got little in the way of human intervention.

Here are the fundamentals:

- Don't let your pigs suffer, but realize that when it is 90 degrees out, and they are sleeping under the juniper trees, they are not suffering — leave them be.
- Never let the pigs go without water. In winter, when temperatures are low enough to freeze the water, offer a sufficient quantity that will remain open for about 20 minutes (or enough time that each pig can become sated) twice a day.
- In the heat of summer, your pigs need shade (white-haired pigs can sunburn) and a mister, sprinkler, or puddle to help them cool down. A mud puddle is free — except for the cost of water — and the mud coating helps keep flies and other annoyances off the animals.
- Year-round, but especially in cold seasons, supply the pigs with plenty of hay or straw; if the material is 2–3 years old, you can buy it for next to nothing. Your pigs will use it to nest and make tons of garden mulch for you in the process.
- Never beat, kick, or otherwise physically stress your pigs. They are sensitive. Because they can easily overheat with too much chasing around, exercise your position as a "smarter" beast and use treats and the like to outwit them. If you can't get the pigs moved to the new field today, try again tomorrow, and give yourself a break too.
- Even on lush pasture, offer your pigs some feed in the form of grain or pellets. And of course, always give them appropriate household or other food scraps. Pigs are not prone to overeating, and when you provide them with ten pounds of lamb's quarters (a robust annual weed) each, they'll leave most of their grain uneaten. Provide minerals as recommended for pigs *on the ground,* not in confinement.
- Spend time with your hogs. Watch them, listen to them, scratch them

Fig. 2.8: *Eight feet in the feed trough and four more on their way.*
Credit: Karen K. Will

Fig. 2.9: *Mud is a must whether you have two hogs or twenty. If you are concerned with permanent damage to the ground, create a feeding/watering/ wallowing area and let your hogs enjoy it.* CREDIT: KAREN K. WILL

behind the ears from time to time. They will learn to trust you, and you will learn to distinguish normal from abnormal. And you just might find a compelling form of entertainment in the process.

- Think carefully about parasite and medication programs, and implement as you are comfortable doing so. Some states have minimum requirements for medication if you plan to sell breeding stock. Use common sense though — if your animals are growing poorly, you might try de-worming them — but it might be that they are simply a slow-growing breed. If they are doing fine, don't feel like you have to meddle.
- Finding a large-animal veterinarian who understands and respects your management strategy is key to your success and enjoyment. Ask your average industrial farm vet about meds, and you will wither at the prescribed program, never mind the unnecessary cost. However, if it's clear that your pig is suffering, don't hesitate to intervene with an antibiotic after a call to your vet — why let the animal remain in pain when you can fix it? This is one of the most controversial areas of pig husbandry. Read all you can on all sides of the medication/parasite control issue before drawing your own boundary lines.
- Have fun. When you are happy, your pigs will be happy, and they will reward you with entertainment, sustenance, and all kinds of valuable farm labor.

On Breeding

As with many animal projects, the temptation will be there from the get-go to breed your pigs. Breeding requires considerably more care and management — you will need a boar and (ideally) separate quarters for him when he's not in use. Likewise, managing a group of pregnant females takes a little more in the way of observation, and each gilt or sow will want a fairly private and safe place in which to give birth and nurture her babies through the first couple weeks of life.

Spend a few years with a non-breeding group of hogs and study up on breeding and birthing strategies before you take the plunge. A visit with a more experienced swineherd who raises pigs on pasture will do wonders to help you decide if that route is the right one for you. Some folks just toss a boar and a few sows out on pasture and let things sort out the way they will. But, if you do your homework ahead of time — even just a bit of research — we believe the experience will be safer and more rewarding for you. Who knows, you might even create a small and profitable business for yourself in the process.

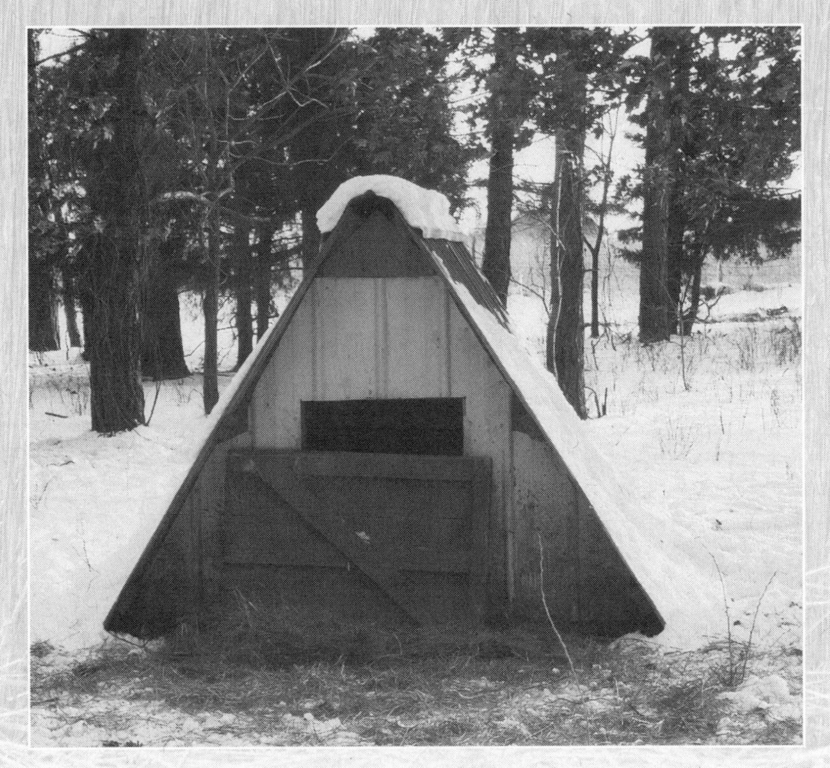

Fig. 2.10: *There's a sow inside this A-frame hut with a litter of babies. She stuffed the hut with hay and plugged the door with hay the night she gave birth. Once she opened the entrance, the authors blocked the door after finding some of her babies wandering around in the snow.*
CREDIT: KAREN K. WILL

Chapter 3

Collecting Solar Energy with Ruminants

Once part of most diversified and subsistence farms, four-stomached ungulates such as cattle, sheep, and goats are designed to harvest the solar energy that's already been captured by plants. It is because they eat that converted solar energy to use as fuel for themselves that they can provide us with such a plethora of products and services. Farmers and homesteaders of old knew that the road to real wealth was paved with photons, neatly converted to food, fiber, fertilizer and mechanical work, and heat (at one time, many northern farmhouses were attached to the barn so the animals' body heat could heat the house). So important was the harvesting of light energy through animals that no farm could be without them.

As modern agriculture became industrialized, it also became less diversified. Farmers started specializing in just a few crops or one kind of animal. And the ecological balance on the farm has been way out of whack ever since. When animals are moved off the land and into the prison-like torture chambers we currently call feeding operations, the animal benefit to the land, the farmer's mental health and well-being, and the environment are lost. Perhaps most unfortunately, any semblance of self-sustainability was lost the moment agriculturists were brainwashed into adopting these methods. One of the principal flaws with current industrialized agricultural models is that it requires vast quantities of a finite energy source (petroleum) — ironically captured from the sun — through the

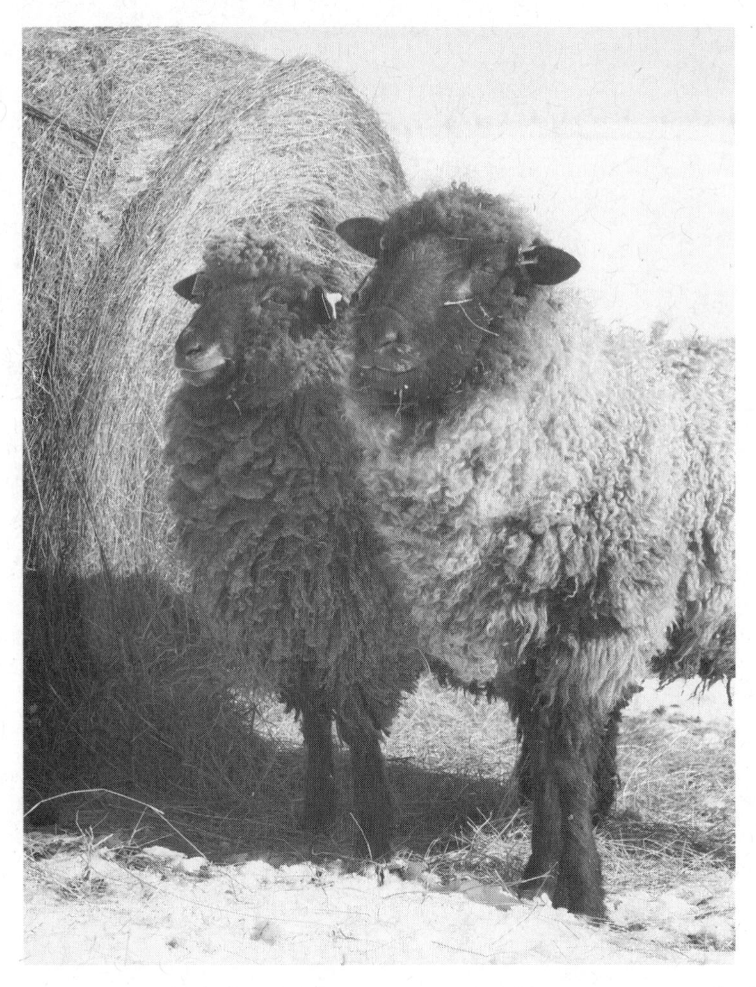

Fig. 3.1: *All it takes to winter over ungulates is plenty of fuel in the form of hay and shelter from the wind. Contrary to popular belief, a barn is not often required.* CREDIT: KAREN K. WILL

very photosynthesis process that only partially fuels industrial ag — and stored in rich and highly concentrated deposits millions or more years ago.

Manage for the Ecosystem You Want

We've discussed the incredible power of hogs for tilling the ground and removing stumps and the like. But what do you do if you want to convert or restore an overgrown scrubland into productive grassland? You could opt for some quality time in the seat of a large diesel-fuel-guzzling machine and doze, shred, or otherwise destroy that unwanted vegetation. But, though operating heavy equipment can be fun, it is hard on a body and hard on the land, and it's noisy, smelly, and pretty expensive. It's also quick. But the 21st-century homesteader has nothing, if not time. Your place won't be built in a day, so take it one day at a time and marvel at your progress — employing a progression of animals to restore (and later maintain) a grassland can be soothing rather than jarring, and you'll end up with something useful even as your land is being restored.

Likewise, if you wish to open up the understory in a woodlot, or clear heavy growth between rows of fruit trees or cultivated vines, well-managed groups of animals can help you make it all happen, and you won't need to spend a dime on feed or fuel. In fact, you may well add value through increased animal numbers and/or through the products they offer while making a living from the growth you're trying to get rid of.

Goats Are a Good Start

When faced with an impenetrable thicket, whether in a meadow, wooded area, or fencerow, nothing beats a goat's agility and tenacity. They will eat a path through almost any landscape. Goats are designed to browse, that is, graze off the ground on leaves and fleshy twigs of woody vegetation. And, if you have allowed your goats to keep their horns, they'll not only eat holes into the thicket, they'll also hook and break stems and saplings by rubbing their heads on them. Goats have been used successfully to keep power line rights-of-way clear of shrubby growth (including tree seedlings) and to eradicate the rogue multiflora rose bushes our USDA once imported and encouraged us to plant in the hedgerow. Goats have the added advantage of being able to go where no bulldozer or tractor could be safely operated.

If the shrubby areas you wish to open up are small, you might want to fence the entire area for goats. If your fencing is goat-tight, it's also pretty much impervious to other forms of livestock. Or, if you're just working on small patches of scrub or rose bushes, you might opt for temporary electric netting fence — assuming you've trained your goats to respect it. We've used electric net fencing and small flocks of Pygmy goats to remove large multiflora rose infestations — all you need to do is surround the rose thickets with the fencing, then let the goats do their thing. Since Pygmy goats are short, after a few days we knock down the taller bushes with a machete — if you keep at it, you'll only need the machete once. You might only get one tenth of an acre macheted in a day, but keep coming back the next day until it's done. You can let the goats

Talking Goats

- Kid: young caprid — newborn through about 1 year
- Doe: mature female
- Doeling: young immature female
- Buck: mature male
- Buckling: young immature male
- Wether: castrated male

Fig. 3.2: *Even dairy goats can work wonders for opening up woodland understory and keeping woody species in check. Note the guard dog — a member of the herd whose job is to protect it from predators.*
CREDIT: KAREN K. WILL

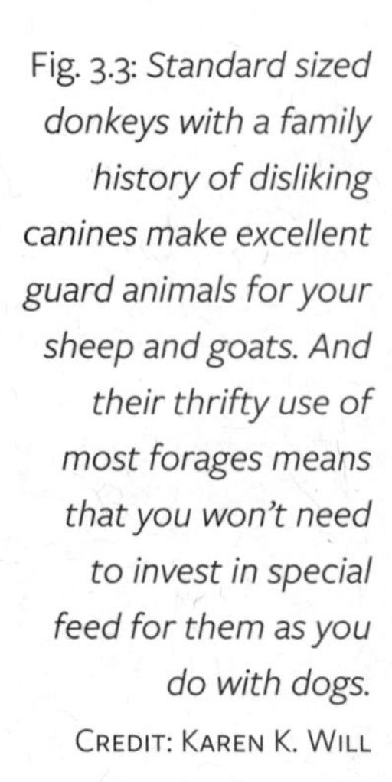

Fig. 3.3: *Standard sized donkeys with a family history of disliking canines make excellent guard animals for your sheep and goats. And their thrifty use of most forages means that you won't need to invest in special feed for them as you do with dogs.*
CREDIT: KAREN K. WILL

re-browse the sucker leaves that will inevitably sprout, or you can move them on and bring in a few sheep.

Goats can thrive on a steady supply of shrubby browse, and they'll create more goats on such a diet if you let them, which makes the exercise of eradicating unwanted vegetation a potential paying proposition. If you enjoy eating goat, you can also feed your family — and perhaps several other families in the process. But don't try to run a goat dairy on this kind of forage. The logistics and fluctuations in browse quality would make dependable milk production tough, if not impossible, to achieve. However, if you do run dairy goats on your pastures, they may perform stalwart service in keeping brush at bay — unless they are hopelessly spoiled by prime hay, lush legume pastures, or other high-value feeds.

If you bring your goats into a corral or barn each evening, you'll also benefit from a good supply of fertilizer. If you don't bring them in, no worries; they will simply convert vegetation to fertilizer and drop it where they harvested it. In this way, you can actually help enrich somewhat depleted areas or jumpstart some of the nutrient cycles by speeding up the decomposition process.

Goats, especially smaller breeds, hornless breeds, or dehorned individuals, will require plenty of protection from predators. When enclosed in electric net fencing, they should be pretty safe, but when out on range, even with good woven wire or electric fences, proven guard animals in the form of dogs, mules, donkeys, or llamas will lend more protection. If you have pet dogs, be aware that your guard animals will likely see them as predators, so don't allow the pets to mingle too closely with the guards, or you might end up heartbroken.

Talking Sheep

- Lamb: young ovis — newborn through about 1 year of age
- Ewe: mature female
- Ewe lamb: young female
- Ram: mature male
- Ram lamb: young male
- Wether: castrated male

Let the Sheep Do the Shearing

Sheep enjoy browsing and grazing, and they'll relish the sucker growth from the shrubs that the goats beat back to the ground. They'll love to browse lush pastures, but also the coarser grasses and forbs that finicky eaters like cattle often ignore. If you follow goats with sheep (and allow them access several times during the growing season), they will quickly kill invading shrubs as they browse the sucker growth. Sheep are quite simply made for keeping woodlot understories open and grassy meadows free of shrubby growth.

When the great cattle barons of the American West felt threatened by sheep, people and animals not only lost their lives, but the sheep were blamed for every rangeland degradation issue imaginable. Sheep were said to overgraze, draw coyotes and other predators, cause soil erosion, cause compaction, spread weeds, you name it. The fact of the matter is that sheep, like all ungulates, will damage a rangeland ecosystem when space is limited and animal numbers are high — the same as all grazing animals.

Sheep do have the ability to clip grasses and other pasture plants very close to the ground. If allowed to overgraze any large area, the entire stand is left at a disadvantage because regrowth will be slow and fewer grass plants will regrow because their growing points — the crowns — are damaged by close cropping. In a fragile natural habitat, this damage can be severe to the point of erosion, invasive weed infestation, and even desertification. But the fault doesn't lie with the sheep; it's largely due to the lazy management practices of the shepherd.

When few enough head are given a large enough space, they will roam from place to place and rarely hit the same patch of pasture more than a couple of times in an entire season (assuming water sources are well-distributed). Likewise, while the sheep's sharply pointed hoof can be a tool of compaction, it can also be a tool of aeration. Hooves readily break up dead, dry, brown vegetative thatch that's unlikely to break down without a little mechanical help. A small group of sheep can help keep open areas open, build soil organic matter, and maintain plant and insect species diversity in meadows, all while providing sheep, wool, meat, fertilizer, and so on. Even if you skip the meat and fiber, sheep make excellent working partners — and they're generally easier to handle and less intimidating to most folks than goats.

Sheep, even more than goats, will require protection from predators like coyotes and domesticated dogs. Luckily, they aren't quite the escape artists goats are, so fencing for sheep needn't be quite so sturdy. Sheep have a very strong flocking instinct compared with other grazing groups; as individuals, they tend to give up more readily (read: they die easily). In our hands, sheep go into shock after relatively minor injuries (what appear to us to be superficial scratches), especially at the jaws of a dog, and they die if left unattended. For the smallholder, an easy way to

Four-Legged Lawnmowers

Sheep have been maintaining country manor lawns since lawns were invented. Their ability to crop the grass short and their penchant for forb-based forage makes them perfect for managing the homestead lawn. All you need to make it work is a sufficiently sheep-proof fence to keep the animals where you want them and sufficient sheep to keep the lawn clipped, yet lush.

We've used sheep to mow the grass for years, and it really works. We fence them away from young trees, shrubs, the perennial gardens, vegetable garden, and places where we don't appreciate their pellets underfoot. We use a combination of woven wire fencing and portable electric net fencing to guide the animals, and we never let them have access to the entire yard at the same time. So a group of ten ewes and their lambs might trim our side yard in two or three days, then we move them on to the backyard, and so on. We don't use any chemical fertilizers or pesticides on the lawn — you shouldn't either, if you're going to mow with sheep.

The first time we tried mowing with sheep, it was because we were caught short of early spring pasture. Faced with the last of the hay fed out and a semi-formal lawn looking shaggy, we parked the petroleum, left the tractor in the barn, and led the sheep to the lawn. Not only did the yard hold the sheep until we got some new pasture fencing in place, it proved to us that putting the concept into practice was really a piece of cake. Generally, we avoid grazing the lawn when it's very wet. Most years, we run the yard mower twice or three times, at most.

deal with predators is to simply bring the sheep in at night. If you have strong coyote or domesticated dog pressure, pen them in a corral fenced with dog-proof woven wire or reinforced with electric fence. You can also pen sheep up in a shed for the night. If you choose this method, be sure that the ventilation is very good. Dust coupled with damp animal breath is a good recipe for passing respiratory pathogens through a flock.

Cattle: Not Necessarily King of the Homestead

The family cow might be third in line behind the barn and tractor as rural icons, and cattle most definitely have a place on the 21st-century homestead. However, unless you have experience with cattle or a true burning desire to own them, don't start with cattle. The reasons are simple: cattle are large, can be intimidating, and (usually) require much more in the way of facilities than other ungulates.

Cattle are excellent grazers and decent browsers, so you can use these majestic animals to help maintain your meadows and woodlands — at least where you want a park-like, open understory. Cattle are often grazed below pecan trees or in orchards using careful management to prevent them from damaging the trees through rubbing, over-browsing, or compacting the soil above the root zone. Cattle are quite selective in their grazing and will leave coarse, heavy growth for later — or they'll avoid it completely. (If this is a problem, running sheep after the cattle can help you manage to harvest even the coarse material without resorting to mechanization.)

If cattle are your only ungulates, then it will be important to manage their access to specific areas of the meadow and the length of time they have access. If you don't, plant species diversity will suffer, and you'll be inviting weedy species to take hold. Some cattle breeds are useful for managing invasive woody species such as eastern red cedar (on the Plains) and for keeping the understory open. Horned breeds like Highland and Pineywoods cattle will rub small cedars to death with their horns, and they'll prune the lower branches of evergreen and deciduous trees alike through a combination of browse, head scratching, and rubbing. Turn a herd of Highland cattle into a dense grove of cedar, and soon

Talking Cattle

- Cow: mature female
- Bull: mature male
- Calf: immature bovine
- Steer: castrated male
- Heifer: adolescent female who has not yet given birth
- Head: one individual

Fig. 3.4: *Give them sufficient time, and Highland cattle in Kansas will take care of any rogue eastern red cedar trees invading the pastures.* Credit: Karen K. Will

Fig. 3.5: *Bulls, just like rams and bucks can be docile and relatively safe to be around. In general though, it's prudent to never turn your back on a bull for too long.* Credit: Karen K. Will

enough, you'll be able to look through the trees as the animals prune the lower branches. Likewise, if an overgrown ravine makes a decent winter camp for your animals, consider feeding them hay down there, and their camping will go a long way toward opening it up.

When you have sufficient grass or other good forage, you can certainly keep a cow for milk, and she'll make you plenty while keeping your pastures in decent condition. Remember that making milk means making babies. If you have no use for a calf — or a bull — then the idealized milk cow might not be the best choice for your place. But if you have your heart set on it, there are ways to manage. Plenty of folks strike up a barter with someone who has a bull — ten pounds of homemade cheese might be all it takes to convince your neighbor to come get your cow and run her with his bull for several weeks. Alternatively, you might trade the calf for bull services if you aren't interested in selling it or raising it for meat or milk.

If you simply wish to have a small herd of cattle to help you maintain the place, steers (castrated males) are a great choice. Relieved of their testosterone-producing testes at a young age, steers can be quiet and relatively gentle. Some folks make a pet out of a steer to help them handle the entire herd more easily. For example, an older, tame steer can be used as a group leader in a herd of young calves — you call the steer in for supper, and the rest of the gang will likely follow. Other folks simply keep a few steers around so they can keep cattle without having to worry about some of the problems that come when cows reproduce. (Bulls can be cantankerous and cycling cows can be quite fractious.) Folks interested in both maintenance and meat often source young steers each spring and send them to market each fall. Some of the finest grassfed beef is produced that way. It's win-win for the farmer and a humane life for the animal.

Social Considerations

Grazing animals, by their very nature, are social. They're genetically predisposed to live in large groups, so they may exhibit some undesirable behaviors if living alone. Three makes an ideal minimum number of animals for a herd or flock because it allows for the formation of a pecking order, which relieves tensions that often arise when there are only two animals in a group. Larger groups offer many social connections, grooming opportunities, and calmer interactions than small groups do. Some of the meanest bulls and cows we've known or heard about spent the bulk of their lives in isolation.

Herd animals generally see humans as predators; however, when babies are bottle fed, they can imprint on humans. This can be useful for gathering the flock or herd, but it can be

Fig. 3.6 *Getting to know the boss cow makes gathering and moving your herd safe and easy. With any breed, nose touches are fine, but you want to avoid scratching them on the poll (where the horns grow), or they may wind up crushing you with their head one day — when they're just asking to be scratched.*

Credit: Oscar H. Will III

a nuisance — or even dangerous — if the bottle baby in question grows up to be an 1,800-pound bull or cow. Even without bottle feeding, if you can convince the lead animal to follow you or come to a feed bucket, you can lead the animals where you want them to go, rather than needing to herd them. In most cases, the herd instinct is sufficiently strong that if you get a group moving in the right direction, the others will follow.

Soil Impact

Grazing animals will work to improve soils in several ways, depending on your management of their movement. In general, when a pasture plant produces above-ground biomass, it also produces a proportional increase in below-ground biomass. So, when the grasses leaf out and grow lush, their roots grow to support the above-ground growth. When the grazing animal (or mower) cruises by and trims back those leaves, the roots die back proportionally. While this might sound like a bad thing, it's a good thing for the plant; the plant wants to keep its nutrient- and water-harvesting structures at the optimal size because it takes energy to maintain them. A just-mowed grass plant with a huge root to feed would surely suffer more than one that can rapidly scale the root back. What becomes of that dead root, you might wonder?

As the roots die back, they leak organic materials into the soil, and, as they decay, they become fodder for beneficial soil invertebrates, whose carbon-skeletal remains will also eventually add organic matter to the soil. So, the very act of grazing puts organic matter into the soil. If you manage your herd or flock so that the grazed plants get a sufficient period of rest, they will regrow and eventually can be grazed again — and yes, they'll deposit another dose of organic material into the soil. But that's not all.

Roots have a marvelous ability to grow through tough soil substrates, and, in some cases, they also penetrate tough subsoils. The passages left in the soil after the roots die and rot away help aerate and facilitate water movement and storage in those soils. So, well-managed grazing intervals will build soil organic matter and positively affect the soil's ability to percolate and store water as well as facilitate more efficient gas exchange. It's an impressive ability. The key to putting it to work for you is to space short-term grazing periods with sufficient recovery time. If you think about it, that's essentially what you're doing when you mow your lawn on a weekly basis.

There's still more, though. Hoof action from grazing animals in a well-managed system will work to break down air- and sun-oxidized plant material at the surface and help work

Fig. 3.7: *Sheep, cattle, and even goats running together will make thorough use of the various classes of forage your place has to offer. Running them separately gives you the opportunity to change the forage matrix to suit your needs.*
CREDIT: OSCAR H. WILL III

it into the soil. This will — yes, you guessed it — add even more organic matter and worm food to the soil. In the process, the hoof action will leave some soil exposed, and it may also press grass or forb seed into contact with that exposed soil. Voilá! You have new plants sprouting where they were once suppressed by thatch.

Grazing in woodlands isn't always beneficial to soils. Woodland soils are typically low in organic matter, and nutrient cycles in those systems work a bit differently. However, using animals to open up the understory can help seed the soil with shade-loving plants that will offer a bit of grazing and help feed the trees.

Significant soil damage can be caused by overgrazing until the ground is bare, running too many head in the woods for too long, or running too many head in an area when the soil is very wet. In the first case, you run the risk of causing erosion, which, on a large-enough scale in a delicate ecosystem (especially one with low rainfall), can lead to desertification. Too many head in the woods can lead to serious compaction, which can starve tree roots of oxygen and cause erosion in some circumstances. Pugging wet soil with deep hoof prints will not only tear up a pasture's plant matrix and physical structure, it will lead to serious compaction, clodding, and other soil structure problems that can take years to repair.

When it comes to grazing finite herds in finite spaces, it is up to you, the land steward and animal husband, to manage for the best. And when you manage for the best, your animals will thrive — and you'll be improving the ecosystem in the process.

Managing Fertility

Most homesteads have sufficient fertility to get most of what needs doing done. Unfortunately, most of the time, the fertility isn't where you need it to be. Grazing animals can help move nutrients from areas where they are high to areas where they are relatively low. They can also move fertility to a central stockpile; then you can spread it precisely where it's needed.

Does the grass grow greener and lusher in that low, where runoff has been accumulating for decades? If so, you can move the fertility back up the hill by grazing the low for a period and then moving the animals elsewhere before they eliminate. Sure, some pellets or pies will fall right back on the ground in the low, but if you time it right, the bulk of the good stuff will get deposited in the next paddock. Likewise, if you want to move some of that fertility from the pasture to the vegetable garden, you might give the sheep the low for a few hours and then bring them to the corral or barn for a bit of thoughtful rumination. You will need to collect the bounty and apply it to the garden, but it will get there eventually.

If you make hay for your animals but don't want to deplete the soils on your hay ground, then feed the hay back to the animals on those paddocks. Or, if you make hay in that lush low this year, you can feed it to the animals where you want the fertility transferred over the winter. As with so many aspects of 21st-century homesteading, this method of manure management takes time, but the end result is rewarding. And it doesn't cost a dime in machinery.

Critics of these types of husbandry methods often cite the need to spend "too much" time with the animals as a major downside. We'd bet those critics have never experienced the joys of spending time with livestock. For us, spending time enjoying the animals and watching them do what they were designed to do is compelling.

Mechanical Work

Our ancestors counted on four-stomached grazing animals for sheer mechanical work, and some folks still do today. With sufficient training (of both animal and handler), goats and cattle can become an efficient team for handling much of the work that is done with a tractor.

Consider this scenario: You keep a small herd of steers (to maintain your land), two of which are sufficiently matched to make a good oxen team. Fast forward through the training, and your meadow-maintaining animals are now capable of taking on further drayage duties — all for the price of some good pasture, some rigging, and some patience on your part. Their feed bill might go up a bit if you use them with a land plow to break sod for days on end or to quarry sufficient rock to build your dream house, but you would have to search far and wide before finding a path more satisfying to walk than partnering with animals in this way.

A note of caution — it takes years to develop good teamster skills, and developing great teams takes time. But if you commit yourself to the task at hand and stick with it, the reward will be yours. To gain experience, many people begin with goats and a goat cart. However, even though goats are smaller and less intimidating than cattle, don't expect to succeed overnight.

Fig. 3.8: *Keeping a ram (center) is necessary when breeding sheep is part of your plan. Remember that even a friendly ram can be dangerous when ewes or ewe lambs are in season.*

Credit: Karen K. Will

To Breed or Not To Breed

Spend a little time visiting online discussion boards or reading the mounds of papers published on animal care, and it's easy to conclude that keeping any intact male animal around your place is a disaster waiting to happen. One expert cautions you to avoid jack donkeys under any circumstances; another says bulls will kill you; and still others caution that a buck with horns cannot be tolerated. While male ungulates are certainly a force to be reckoned with, they are neither scary, nor necessarily deadly — but they are unpredictable, especially if they view you as competition when females of their species are in season. All of the things that folks say will happen could happen, but at some point you just might want to stare those overblown specters down. Especially if you have to, in order to accomplish what you want to accomplish.

Rams, bucks, and bulls are generally not to be trusted, and we'd never recommend making a pet of one, but if you want to breed your animals and don't want to fool with artificial insemination or hauling your females to a male (or a male to your females), you'll just have to face it.

The first step to breeding is having a plan, and the second is readying your facilities and checking your skills.

Fig. 3.9: *Some ewes routinely give birth to triplets and raise them, however it's prudent to watch to ensure that the runt gets sufficient nutrition early on.* CREDIT: KAREN K. WILL

Fig. 3.10: *When there's a triplet who doesn't get its fair share, or you wind up with orphans for any other reason, bottle feeding a few lambs will make them "people friendly" and easier to handle as adults.* CREDIT: OSCAR H. WILL III

When you bring males in, you will need to separate them from the females until you're ready to begin the process. This is especially true if you want to time lambing, kidding, or calving for a specific month or season of the year. Some folks don't worry about timing too much, but you still want to be able to separate off the males, just in case they harass the females during, before, or after they go into labor.

If you plan to be relatively organized about this, it will be useful to have a couple of other animals on hand that your ram, bull, or buck can hang with during the months when he's not "working." Some mixed-breed farms and ranches report great success with inter-species groups. For example, rather than penning your ram alone except during the breeding season, you might run him with other species. We have several rams, and they run with the male donkey or cattle for most of the year. So strong is their flock instinct, that once their breeding duties are completed, the rams will actively seek out their old companions. We also had a lone goat that liked to sleep in a pen he shared with some poultry; during the day, he liked to run with the sheep.

Now that the issues with the breeding males are worked out, what about the male offspring? Another reason you'll want sufficient facilities to separate males from females is that at some

Fig. 3.11: *Calves are cute and fun to watch — but don't mess with junior too much unless you want to have a close encounter with big mama.*
CREDIT: KAREN K. WILL

point little bull calves, ram lambs, and buck kids will start producing testosterone and other androgens. When they do, they'll pick up some bad habits — like harassing their mothers, sisters, and virtually any other female member of their species. However, bad behavior notwithstanding, "baby" boys and their moms dislike being separated. They'll find every fencing malfunction on your place — or quite possibly create a few of their own — just to get back together. This is the weaning process, and it is often a time of pandemonium.

Our farm is home to not only intact boars (discussed in Chapter 2), but also intact bulls and rams. Though rams may try to ram you if they are cornered (and the herd bull's horns measure about four feet tip to tip!), if you don't let yourself trust them and expect the worst of them most of the time, you'll anticipate and avoid danger.

So if you want to try breeding, by all means give it a shot. Read all you can about the species in question, talk to level-headed folks, select breeding stock that suits your purpose, take a deep breath, and remember that people have been doing this for thousands of years, and the animals have too.

Dogs and Cats Will Work for Food

Shelter dogs are loving, smart, and deserving of the best homes. But purpose-bred dogs were developed (similarly to livestock) to do particular jobs. Dogs with working parents bred to herd or to keep vermin at bay will be much more likely to perform those tasks naturally than pet- and champion-grade purebreds or mixed breeds from the shelter. We love all dogs, but we rely on our farm dogs to do real work.

Among our favorite purpose-bred dogs are those that hail from the border territory between England and Scotland, particularly the terriers and collies originally bred for work and companionship. Our border collie, Clover, comes from working parents, and she's smart, serious about work, and aims to please. Her job is to round up loose critters like poultry and make life uncomfortable for the local coyote pack, possum den, and raccoon lodge at night. Border collies live to work, and if you don't offer them a job, they'll invent their own — including harassing your sheep to death. Training is a must.

Fig. 3.12: Credit: Karen K. Will

Our two terriers — Molly, a border terrier, and George, a Cairn terrier — are robust and industrious little work dogs. They surely believe they're much larger than they actually are when encountering marauding bands of raccoons, and a possum playing possum doesn't fool these dogs one bit. These terriers were bred to be good mousers and ratters and will dig their prey from holes in the ground and rout them from rock piles, brush piles, and rock walls. They're up for long hikes in any season, love to go for a ride in the truck (in the cab, of course), and are generally up for a swim in the pond any time. Terriers will run off and chase the ghost of a critter in the blink of an eye, so it's very important to keep these dogs close at hand while on the farm. At the end of the day, they'll sit in your lap as you read, or make themselves comfortable stretched out by the fire. Terriers need you to take charge. Make it clear that you are the top dog in the pack. As with all dogs, training is critical.

In our experience, these small terriers are better mousers — both in the house and the barn — than any cat we've ever had. We do keep a cat in each of our major outbuildings for rodent control, however. The best cats we've had for rodent control all came from farms where they were born in the barn to mothers and fathers who were largely responsible for feeding themselves. You needn't worry about wiping out the local songbird population with barn cats; they find rodents to be much easier prey.

Section 2

Bounty in the Woodlot and Hedgerow

Some modern farmers look at an overgrown fence line or hedgerow as a sign of poor management, or even worse, a waste of valuable farm ground that could be devoted to cash crops. It's true that trees and shrubs growing along a fence line can become a nuisance when it comes to keeping the wire tight and the animals in, but with a little care and routine maintenance, the hedgerow's value to wildlife and its provision of wind protection, shade, and privacy screening far outweigh any negatives. Likewise, that wooded draw in the west meadow or east pasture can offer you so much more than a place to park your bird-watching blind or a way to provide your grazing animals some shade.

Your hedgerows, wooded acres, and overgrown creek beds hold aesthetic and environmental value, which is pretty good by itself. But, they are also great renewable raw material suppliers. When it comes to raw material, most folks think of the forest as fuel, but, on the small homestead (or even in town) hedgerow saplings, woody shrubs, and trees can provide you with a seemingly endless supply of material for creating everything from wooden spoons and furniture to animal shelters and raised garden beds. Once you start thinking about the woody growth around your place in this way, you'll plan your cutting (and planting) more carefully.

In the following three chapters, we consider ways that you might put your woodlot to use in the creation of sturdy fences, structures, and tools that will benefit your place and your pocketbook.

Chapter 4

Fencing and Fences

Good fences make good neighbors — no doubt about it. Nothing can make your homesteading experience go sour faster than making enemies of the folks you may need to count on the most. Since fencing is one of the more labor intensive and expensive components of building a homestead, it often gets neglected, but it doesn't need to be that way.

Folks build fences to keep their animals in, others' animals out, for aesthetic reasons, for security, to help manage weather, you name it. If a trip to the local farm supply store for fencing supplies has you feeling faint, try taking a look around your place. Soon enough, you'll learn to recognize the fencing resources you have right under your nose.

Living Fences

When exploring ways to put a woodlot to good use, it's easy to forget that our ancestors actually put it to use in living form as a fence or hedgerow. Folks tended to settle their homesteads early in life with the notion of staying there forever, so they'd collect seeds and/or seedlings from the woodlot (or along creek banks) and plant them in dense rows as protection from wicked winter winds or as enclosures for livestock. In many places, natural materials were the only materials available for building fences of virtually any kind.

Although an animal-containing living fence can also serve as a windbreak, more effective windbreaks can be created using multiple rows of trees and shrubs that grow to varying heights.

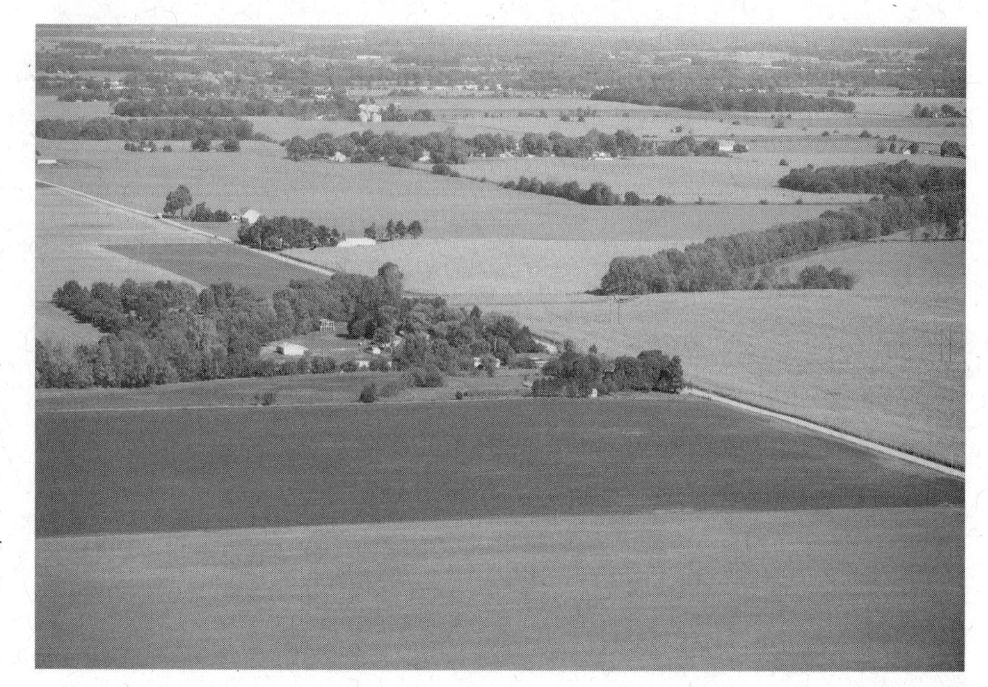

Fig. 4.1: *Living fences in the form of hedgerows and shelterbelts mark the locations of building sites across much of the rural Midwest and Great Plains.* Credit: Oscar H. Will III

In much of North America, true homestead shelterbelts or windbreaks weren't planted on a large scale until after the dust bowl, although savvy northern farmers and even the railroad had been planting trees for decades in long, wide rows to block the winter winds and allow the snow to drift in a controlled manner, away from tracks or critical structures.

When planned carefully, a windbreak or shelterbelt and a livestock-containing hedgerow can be managed for fuel, wildlife, forage, and shelter. If you make the inner row of the shelterbelt a true hedge, the arrangement will do double-duty because it will also contain livestock. For the sake of simplicity, we'll treat the shelterbelt separately from the hedge, but if it makes sense, or if you just want to experiment, you should feel free to plant them in an integrated manner.

Windbreaks and Shelterbelts

The most effective windbreaks to protect building sites and provide good wildlife habitat require at least ten rows of trees and shrubs. In general, windbreaks should be planted with the leeward rows at least 50 feet from any structures you want to protect, and the rows should be placed between the area to be protected and the prevailing winds — winter winds, especially. In the north central portions of the United States, shelterbelts are most often planted on the north and west sides of the protected area.

The classic shelterbelt uses a combination of width and height (in the form of evergreens, deciduous trees, and shrubs) to foil the wind, trap snow, provide wildlife shelter. Shelterbelts can help keep your place cooler in the summer by shading mid to late afternoon sun, and they can trim up to 30 percent off your winter heating bill/ firewood needs. While you might be tempted to plant a shelterbelt all the way around your home, bear in mind that a mild breeze can make life much more bearable in the heat of summer. Although there are about as many models for shelterbelt layout as there are shelterbelts, the general concept is as follows.

Evergreens are used on the inside rows of a shelterbelt (closest to the house or outbuildings) to provide a final wall from the wind, to trap any remaining snow, and to provide snug winter shelter for wildlife. Beginning at least 50 feet away from buildings (and toward the source of prevailing winds), plant up to four rows of evergreen trees — you can mix and match species you have available locally, but you might plant the shorter species such as eastern red cedar in the inner-most row, with taller pines and spruces filling rows two through four. Leave enough space between the rows (12–18 feet is reasonable) so you can easily cultivate the soil between them for at least three years.

Fig. 4.2: *About 12 years after Hank planted it, this shelterbelt provides significant protection for a greenhouse and barn in South Dakota.*
Credit: Oscar H. Will III

For the middle of your shelterbelt, plant the tallest deciduous trees in the two rows adjacent the evergreens; put smaller deciduous trees in the next two rows. You can finish out the shelterbelt on the windward side with a couple rows of woody shrubs that are spaced very close in the rows (six feet or less) — some folks also reduce the space between these two rows, but it isn't essential. When you do the math, you will discover that the "ideal" shelterbelt is 120–150 feet wide. If you don't have that much space, feel free to change things around.

When done right, the outer shrub rows of the shelterbelt will work as a snow trap; as the wind hits the shrubs and slows down, blowing snow will drop out. The wind will further slow and then be deflected upwards to as much as 20 times the height of the tallest deciduous trees. This creates a wind vacuum on the inside. The evergreens will muffle sound, provide wildlife shelter, and help keep your place snug during the wickedest of blizzards. They'll also filter dust, odors, and noise in the warmer seasons.

If you live in a windy, relatively tree-free area, you might be able to get some help with shelterbelt design and installation (or a monetary cost-share toward those ends) from your local agricultural services office. You might also check to see whether there is a state or provincial nursery that can supply you with inexpensive seedlings if you can't supply enough of your own. Be aware, however, that signing up for any government program comes with strings.

Twenty-five years ago was the best time to have planted your shelterbelt, but today is the next-best time. Even if you start with small saplings and seedlings, you will reap significant benefit from your planting within about three

Beware Rogue Hedge Species

If you live in the Northeast, you have no doubt heard of the pasture scourge known as the *multiflora rose*. This brambly, thorny rose was imported from Asia as a rootstock for grafted ornamental roses and was later touted by the U.S. Soil Conservation Service as an ideal species for making living fences. Although the multiflora rose can have a place in hedge construction where goats will graze it, in many areas of North America, the plant's uncanny ability to move out of the hedgerow and into the pasture (especially on forest-type soils) has elevated this rose's status to noxious weed. So aggressive is the plant that it can readily exclude native plant species and more desirable exotic species. And ironically, though the multiflora rose might be able to form impenetrable thickets, for the goat, that thicket is nothing less than a candy store with a seemingly limitless supply of candy. (See Chapter 3 for hints on how to rid your place of this pest with goats!)

Even in Kansas and other Plains states where the famed hedgerow tree Osage orange grows more-or-less natively, the tree can become a pasture nuisance if you don't practice some kind of clipping after your grazing animals are through or if you don't use any heavy browsers like goats in your rotations. In some parts of the country, the imported honeysuckles and even eastern red cedar (*Juniperus virginiana*) can be invasive. Bottom line with hedgerow building is that as you're creating wonderful wildlife habitat, you need to keep in mind that seeds from some of the hedgerow species might get spread where you (or your neighbors) don't want them to grow. Still, the joys associated with a truly animal-tight hedge more than outweigh keeping the less-roguish species in check.

years, and the benefits will increase as the decades pass.

Bull Strong and Hog Tight

No doubt, agriculturists have been using live fencing for many centuries, including fencing to contain animals of one sort or another. In its most modern iteration, a living fence is a fairly thick hedge of trees and shrubs to which barbed wire is attached (and even in-grown through the trees' trunks). While this wire within the structure serves its purpose, a living fence that is free of wire or other metal will not interfere with trimming, harvesting fence posts, or collecting firewood in the future. It will take a little more time and a bit more management to establish a live, virtually impenetrable hedge, but the end product is arguably more aesthetically pleasing, costs less, and won't stress the plants or your chainsaw down the road.

If you have more tree seedlings and time than money, you can create a lovely animal-tight hedge over the course of a few years. You'll need to figure out some way to keep your grazing and rooting animals away from the hedge until it becomes well established. Before you start, though, it's useful to understand a little about how animal fencing works.

Animal fences work in two basic ways: physical and psychological. A 12-foot-tall stone wall in good repair will keep most animals in or out no matter how much they rub, scratch, or try to climb it. Conversely, a fence created with a single electrified strand of lightweight wire offers little in the way of a physical barrier, but it will serve as a psychological barrier once your animals have been shocked by it. The best fences integrate both physical and psychological components — the hedge is no different.

When constructed carefully, with appropriate tree and/or shrub species, the hedge will be physically impervious to most classes of livestock, at least for a while. If at least one of the plant species in the hedge happens to be thorny, animals that lean too heavily on the hedge will get pricked and thus learn to avoid it. However, this only works if the animals are not starved or frightened into wanting or needing to escape. Even cows well trained to high-powered electric fences will readily bear the shock to reach lush green forage when their pasture is depleted. But if they have plenty of good green forage on their side of the fence, they won't lean on the fence just to check out that grass on the other side, even if it is a bit greener.

Now comes the fun part. First you'll want to lay out your hedge, mow the lines close, and turn a furrow (or otherwise till the lines where you will plant the hedge) in the fall. You might employ your electric-fence-trained hogs to tear it up for you using portable electric fencing. The purpose of this step is to allow the turned soil to mellow over the winter. If weeds or grasses begin to sprout, till them down with a disc, tiller, hoe, hog, or whatever you have on hand that meets the scale of the project. Once things go dormant, it's time to consider the plant matrix you'll put together to make the hedge.

For your main structural wood, choose easy-suckering hardwood trees like Osage orange, black locust, holly, honey locust, elms, or oaks. If you have honey locust readily available, use it. If your grove delivers all kinds of Osage orange fruit in the fall, you can't go wrong planting those. Choose a thorny tree, and your single-row hedge will be quite secure. If you plan to purchase seedlings, it's probably best to avoid once-thorny species that have been bred to be thornless. You might not want a thorny locust tree as a specimen in the yard, but its role in the hedge will be important.

You will probably get similar results whether you sow seeds or transplant seedlings; which you choose can depend on whatever is readily available. You'll want sufficient seeds or seedlings to plant one roughly every 12–18 inches in your furrows. Using Osage orange as an example, collect fruits in the fall and allow them to overwinter in buckets or tubs exposed to normal freeze-thaw cycles — even rain and snow. This process will create a fermented mass that may look unappealing, but it will become a bounty in spring. Some modern-day experts suggest that you collect the fruits, store them in a cool place over the winter, and then mush them in 5-gallon buckets by covering with water and waiting a few days until fermentation is just evident (bubbles start to form). Drain and repeat until the hedge apples can be mashed easily.

About the time you would normally plant corn in your area, create a slurry from the Osage orange mash by adding water to the buckets and stirring it all up to break up the pulp. Plant the seeds by dribbling the slurry into the furrow you prepared in the fall. With a little practice, you'll find it possible to distribute the seed pretty evenly along the row. Finally, partially backfill the furrow to cover the seeds — you can press the seeds and soil into contact by walking down the furrow with flat-soled shoes or simply wait for the next hard rain to do it for you.

As your hedge seedlings germinate, you might want to thin and/or redistribute them to get an ideal spacing. If you have space and energy, you can transplant extra seedlings to your nursery plot or an area where you can let them grow for a few years. Even three-inch-diameter trees can be harvested for making fence posts, gates, hurdles, and other farm structures. (Note: You'll only want to do this with naturally rot-resistant species like Osage orange, black locust, or bald cyprus.)

In the fall, plash the seedlings as follows: Carefully lay the seedlings over and, using anchor pins, pin them to the ground about two thirds of the way from the base to the tip. If they're not flexible enough, take a small axe or machete and partially sever the trunk near the base to facilitate laying them over — but make sure to leave at least one third of the trunk intact. Cover the pinned area with soil. Alternatively, this entire procedure can be done by laying the seedlings over in the furrow and pinning them by mounding soil between the root end and the shoot end.

Fig. 4.3: *Layering saplings after their first year of growth will yield a much thicker hedge that's less permeable to livestock near the ground.*
CREDIT: OGDEN PUBLICATIONS, KRISTIN HURLIN

The following spring, the now-horizontal seedlings will send up shoots vertically along the trunk. In the fall, bend these shoots horizontally and weave them together so that you get a woven barrier that's about 24 inches off the ground. In the third year, these stems will begin growing into one another, and they'll also send up thick vertical growth. Weave this growth together in the third fall if you wish, or simply prune the hedge to its final height. Pruning the rapidly growing green shoots several times over the summer will stimulate the lower buds to produce more lateral branches, which will make the hedge even less permeable. For another method of forming the

Fig. 4.4: *Weaving together the second and third year's growth helps make the hedge into a formidable livestock barrier.*
CREDIT: OGDEN PUBLICATIONS, KRISTIN HURLIN

hedge, see below "Caldwell's Preferred Hedging Method."

Living hedges of the kind described above can be created using other species — or even a logical mixture of species. You will need to collect sufficient seed or seedlings of those trees, plant them using best practices, and then weave them as described. In the end, you'll wind up with a lovely hedge that will harbor all manner of insect-eating songbird nests and that will be secure against all but the most determined of livestock and wildlife. With a little time and luck, your efforts will pay with a fence that's bull strong and hog tight.

While single-row livestock-tight hedges are effective, a double row (or more) may give you additional security

Caldwell's Preferred Hedging Method

In his 1870 work, *Caldwell's Treatise on Hedging,* Joseph A. Caldwell notes that the interweaving hedge-forming method may not be the best method for creating truly hog-tight hedges. His method of choice involved mechanically crushing Osage orange fruit and separating seed from pulp with water. He recommends planting dry seed in March, but says that good results can be had with sprouted seed in May. Soak the seed for about 40 hours, changing the water regularly. Mix with sand, place in a box in a warm place, and keep damp until the seed wakes up. Carefully plant the sprouted seed in a nursery plot in drills sufficiently distant from one another that you'll be able to cultivate between them with either a hoe or a horse-drawn cultivator. (Today the choice would more likely be hoe, wheel hoe, or rotary tiller.) In any case, he recommended that the sprouted seeds be placed an inch or more apart in the drill and about an inch and a half deep. Be sure to water as you plant them.

From this nursery, you can prepare your transplanting stock for early winter or spring planting. Prune the first year's growth in August using a scythe or other mechanical pruning tool. Cut the seedlings close to the ground but just above the first node, if possible. Dig them when dormant, bundle the seedlings and bury them in sand or straw — or place in a moist root cellar or box. Caldwell then recommends transplanting the seedlings into the hedge row using the equivalent of a hand tree planter that cuts a slot in the soil for the root (he used a spade) or a furrow cut with a land plow.

Prune the second year's growth in August an inch or two above where you pruned for the first year's growth. And repeat in the third year. The end result of this process is that you get very stout trees with many lateral branches down low and many vertical branches fighting for apical dominance. Keep pruning the hedge at its final height, keep it wider at the base than at the top, and it should be bull strong and hog tight.

and a bit of diversity to boot. With the double-row approach, pick a thicket-forming shrub such as witch hazel, rugosa rose, blackberries or other bramble, or even bush willows. Plant these in a row spaced a couple of feet on either side of your structural row (though putting both on the inside may be most effective for containing species like hogs). And, as with the single-row hedge, by all means mix and match species to suit your needs or situation. When homesteading, it's important to avoid allowing rules to handicap your progress.

If you are at all worried about it, or if you need animal containment right away, you can reinforce your hedge with a single-strand electric fence or a full-fledged field fence, but that might be overkill. If you are short on time, you can go ahead and build an animal-tight field fence while leaving sufficient space to plant your hedge over the course of a few years.

Easy and Inexpensive Fences

Centuries ago, fences became necessary to contain livestock, to keep out predators, and to draw property lines. In the eastern half of the United States, where the soil is rocky, land owners would plow the fields, pick up the rocks, and build stacked-stone walls to fulfill those needs, transforming a nuisance into a resource. They also headed into the abundant forests to harvest the materials needed to craft wooden

How to Grow Osage Orange Trees

In the March 31, 1866 edition of the *Osage County Chronicle*, published in Burlingame, Kansas, the Goodhue & Runnels nursery reported a method for growing an Osage orange hedge that is essentially the same as the one recommended by Caldwell. However, Goodhue & Runnels had this to say about sprouting the seed:

> *First. Sprouting the seed, soak the seed in soft water six or eight days, standing where it will keep warm, changing the water often to prevent fermentation, say every two days, then drain off the water and mix the seed in an equal quantity of sand earth, and let it remain in a warm place, kept moist and stirred once a day until the seed begins to sprout, then sow immediately. If the weather during the soaking of the seed should become too cold or too wet, so it probably would not do to plant the seed by the time it would begin to sprout, remove it to a cooler place so as to retard its sprouting. Plant the seed about good corn planting time, in good mellow, rich, sandy soil. If not sufficiently sandy to keep from baking, make it so by mixing in sand.*

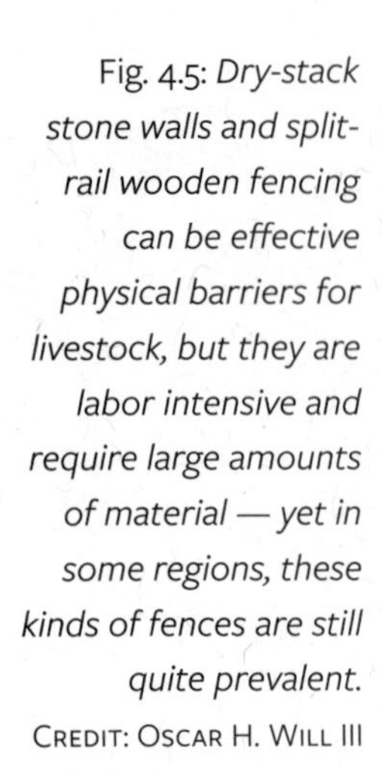

Fig. 4.5: *Dry-stack stone walls and split-rail wooden fencing can be effective physical barriers for livestock, but they are labor intensive and require large amounts of material — yet in some regions, these kinds of fences are still quite prevalent.*
CREDIT: OSCAR H. WILL III

post-and-rail fences around the homestead or barnyard.

A stacked-stone wall built from rocks harvested from your property is beautiful and smart. However, it's fairly technical and labor-intensive work that requires a certain level of expertise — there are books and even associations devoted to the skill. You'll need tons of material (literally), but if done right, a stacked-stone wall will last a lifetime. Stone is particularly useful for building retaining walls, raised garden beds, garden walls, and other landscaping elements, but if you intend to keep your goats in a stone-wall enclosure, it might be best to reinforce the wall with a strand or two of electric fencing on the inside.

Installing a wooden split-rail fence is also a beautiful way to use the natural resources of the farm, though it presents challenges in the form of labor-intensive, precise work and the need for lots of decay-resistant, splittable wood. Back when every family kept a milk cow close to the farmhouse, a simple split-rail wooden fence was just fine for containing a tame cow. With other, more wily and/or aggressive farm animals — like goats or sheep — who rub or push on tenoned rails, this type of fencing fails to prevent animal walkabouts unless paired with some form of wire.

These days, building livestock fence can be as easy as calling a fencing contractor or heading down to the nearest farm supply store and applying some labor. But both of these options are really expensive. And do you really want to spend money on materials if you already have some growing on your place? A rustic wire fence, crafted from fence posts you cut, will look lovely and work every bit as well as a fence made with store-bought components. You'll need to spend some money on wire, tools, hardware and so on, but you can even build your own gates, if you have the mind to.

Fig. 4.6: *Spend a little time with it and you will be able to create lovely and effective fences using posts harvested from your place and some carefully sourced wire.* Credit: Karen K. Will

Lay out the Footprint

To get an accurate measure of the fence's length, you'll need to measure or pace the fence's path. Straight lines will work better than curves, so if you need to fence curved areas, render them into the least number of straight sections you can because all the corners that connect straight sections require brace structures, which are labor and material intensive. Next, count all of the corners and determine where you will want gates. Once you have that information, you can figure the number of posts and length of wire your project will require.

To enclose a 100-foot square (10,000 square feet, slightly less than a quarter acre) and install one corner gate, you'll need five anchor posts (corner and end), eight brace posts, eight brace rails, and about 32 line posts, give or take a couple. For a fence this size, you'd save close to $500 by harvesting the posts from your land rather than buying them from a farm supply store. Enclosing larger areas will often reduce the cost per square foot, especially on a relatively square area. If we expanded the above example to a 150-foot square, the number of line posts and length of wire are all that would increase.

Harvest the Posts

Once you know how many posts you'll need, head out to your woodlot or hedgerows to identify suitable candidates. Select strong, rot-resistant trees with relatively straight trunks like Osage orange, black locust, eastern red cedar, bald cypress, or white oak. With deciduous trees, look for groups of 4- to 8-inch-diameter sucker trunks sprouted from a large stump; harvest them all unless it will make a hole in your windbreak. Or, look along the fencerow for groups of trees that are crowded, and thin them out.

Using a chainsaw (a handsaw will also work), cut line posts from 3- to 4-inch-diameter stock at least 2 feet longer than the desired height of your fence (7-foot-long posts should work for most); corner and brace posts should be in the 5- to 8-inch-diameter range (larger diameters work well and can also be split); cut them at least 3 feet longer (8-feet long works for most) than the desired height of the fence. Cut horizontal brace rails about 8 feet long from 4-inch stock. Grubbing the posts back to your fencing site is much easier if you have access to a pickup truck or tractor. When working with very large diameter or very long trunks, it's often safer to skid the timbers using a log arch attachment on the tractor. Again, exercise caution and be safe, but do what you need to do. (Dragging a heavy log on the ground using a chain attached relatively high on the rear of a tractor or utility vehicle could cause it to do a back flip, and you could die in the process.)

Good Braces Make Good Fences

Travel any country road, and you'll see plenty of fences with sagging wires or mesh so loose that all but the most timid livestock could pass right through. Focus on the ends or corners, and you'll find posts that appear to rise miraculously out of the ground, tilting precariously. Fence failures such as these are commonplace, but most of those sagging wires are caused by an anchor post being pulled from the ground by the tension of the wires themselves.

Well-braced fence ends and corners aren't rocket science, but you need to consider a little physics — or follow the advice of experts — to get them right. Sometimes, 6 inches makes all the difference between a lifelong fence installation and one that fails in a few years.

The typical tensioned-wire fence exerts a minimum of 1,000 to 1,500 pounds of pull on an anchor post. An anchor post serving as a corner post must withstand that much pull in two directions. Soil movement due to temperature and moisture fluctuation and livestock or wildlife collisions with the wire can easily increase the pull to 2,500 pounds or more.

Obviously, the magnitude of these loads is more than an average post can bear for any extended period, but, with a bit of thoughtfully placed bracing, a 6-inch-diameter wooden end post will offer an anchor capable of withstanding temporary pulling loads up to about 6,000 pounds.

Many brace styles have been used successfully around the world. A few rely on an earth anchor or deadman (such as a large rock or log) buried in the ground and wired to the anchor post opposite the pull imposed by the fence. While these are inexpensive and easy to install, they are not effective when the fence follows or ends on property lines because they extend outside the enclosed area (they are also easy to trip on, catch mowers, and injure livestock).

Other brace styles use a combination of vertical posts, horizontal or diagonal rails, and wire to help anchor the fence. These are installed in the fence line between ends or corners, so they are perfect for perimeter fences that follow property lines.

In places where rocks are abundant and/or it is difficult to set posts at least 3 feet into the earth, a number of stone cairn or crib braces can be used — but they extend out of the fence line, so might not be appropriate for property-line fences.

Single-Post Diagonal Brace

Leave it to our friends in New Zealand to come up with an inline fence brace that's attractive, relatively easy to build, and offers economy of materials and labor. This "kiwi brace" is strong enough to anchor a 660-foot-long fence made with up to six stretched wires. The brace employs a single 6-inch-diameter (or greater) anchor post set vertically into the ground a minimum of 36 inches (deeper is better) and a diagonal rail and slider block to keep it upright.

Fig. 4.7: *The kiwi brace relies on a diagonal brace rail tensioned with a wire to the base of the anchor post to resist the fence's strain. Though not as popular in North America as the two-post horizontal brace, when built well, it is nearly as effective.* Credit: Ogden Publications, Nate Skow

Fig. 4.8: *A piece of rebar or other stout metal rod pins the Kiwi brace's diagonal rail to the post.* Credit: Ogden Publications, Nate Skow

Once the anchor post is set, measure the distance down the fence line that's twice the height of your fence's top wire; center a large smooth rock or concrete block on that point. This *slider block* will prevent the brace's diagonal rail from digging into the ground and allows the anchor post to flex when it has to — as the soil's temperature and moisture changes or when the fence experiences an impact.

Next, prepare the diagonal rails. Select a straight 4- or 5-inch-diameter trunk long enough to extend from the center of the slider block to just below top-wire height on the anchor post. Cut the ends diagonally to match those surfaces.

Install this rail by attaching it at the post end with a ⅜-inch-diameter galvanized steel pin. (You can use rebar in a pinch — or some other metal pin you have in your parts bin.) Tension the brace by twisting (with a twist stick) a couple of wraps of 9- or 10-gauge brace wire looped (and stapled) around the base of the anchor post and the end of the rail about 3 inches above the slider block. If the rail has a tendency to move out of the fence line under tension, drive a stout stake on either side of it to keep centered.

Two-post Horizontal Brace

The two-post brace with a single horizontal rail is considered to be the easiest anchor, and so the easiest to get right the first time. Like the kiwi brace, the two-post horizontal brace makes a good anchor for 660 feet of fence constructed with up to six tensioned wires.

This brace consists of a 6-inch-diameter (or greater) anchor post, a 4-inch-diameter (or larger) rail, and a 6-inch-diameter brace post. Drive or set the posts a minimum of 36 inches into the ground — spaced approximately

Fig. 4.9: *This corner brace assembly consists of a pair of two-post horizontal braces that share a single anchor post between them.* CREDIT: KAREN K. WILL

twice the fence's height from one another. Pin the trimmed rail between the posts just below the top wire (lower, if the rail is shorter than twice the fence height, but never less than half the fence height from the ground).

Tension the brace by twisting a couple of wraps of 9- or 10-gauge brace wire looped (and stapled) from rail height on the brace post to about 3 inches above the ground on the anchor post. Don't over-tension this brace, or it will have a tendency to jack the anchor post out of its hole.

Fig. 4.10: *The rock cairn or crib brace is particularly effective in situations where brace posts cannot be planted in sufficiently deep holes and rocks are plentiful.*
Credit: Ogden Publications, Nate Skow

Rock Crib Brace

The rock crib anchor is often overlooked, even in places where there's plenty of raw material. If you plan to use this brace on the property line, check with your neighbor(s) to be sure they don't mind because if they want to install a fence, they will need to use some of the same anchor posts that you do.

The rock crib brace consists of two or four posts and sufficient wire mesh to form a 4-foot-diameter crib at least 4 feet tall. For end braces, simply install a pair of 6-inch-diameter posts a minimum of 3 feet into the ground (if possible) spaced 4 feet apart. Place the wire-mesh crib between the posts, wire the posts together with three evenly spaced loops of 10-gauge brace wire (these pass through the crib) and fill the crib with rocks.

Take some care with stacking the rocks. You want to achieve the most stable brace you can. The most effective way to brace a corner using this method is to set a second pair of posts at right angles to the first.

Now Let's Build a Fence

Begin by building braces — this example is specific to two-post horizontal braces. Take care to mark the exact locations of the fence's corners, ends, and gate — anchor posts will be set in these locations. Using a post-hole auger mounted to the tractor's 3-point hitch, a handheld power auger, hand-powered auger, or clamshell digger (a "jobber," in some regions), carefully bore holes at least 3 feet deep and place the anchor posts into position.

Fig. 4.11: *One of many variations on the two-post horizontal brace theme employs a diagonal brace. Note that this brace was installed to support strain pulling from the right side of the photo.*
CREDIT: OSCAR H. WILL III

Fig. 4.12: *Wire stretching tools come in many different forms. This antique works on a simple lever system and is still on the job.*
CREDIT: OSCAR H. WILL III

(A tractor's loader bucket and a chain would come in handy here for very large posts, but setting them by hand is totally doable). With the posts braced plumb or with a slight lean away from the direction that the wire will pull, backfill the hole about 2 inches at a time. Take care to completely pack each 2-inch layer with a metal rod or stick to exclude any air pockets or areas of loose soil. The anchor posts should not wiggle if you set them right.

Using mason's string or light wire attached between anchor posts as a guide, bore another 3-foot-deep hole at a distance from the corner post that's approximately 3 inches shorter than the length of your horizontal brace rail. Insert the brace post in the hole and backfill partially; trim, fit, and pin the horizontal rail in place and then finish backfilling the brace post. Complete the brace by securing it with a diagonal twist of brace wire as described above.

Depending on the animals you wish to contain, choose from among woven wire, smooth wire, or barbed wire. If installing barbed or smooth, determine how many strands you will need ahead of time. Wire prices fluctuate greatly, but in mid 2012, 48-inch-tall non-climb woven wire cost about $1.35 per linear foot; a five-strand barbed wire version cost about 30 cents per linear foot.

To install the wire, first attach it to a corner or end post; use a combination of fencing staples and the wire itself wrapped around the post and twisted onto itself. Pay out sufficient wire to extend at least 2 feet past the opposite corner or end post, and attach it to one end of a wire stretcher. Chain the other end of the wire stretcher to something solid — your tractor's loader bucket, your pickup's trailer hitch, a nearby tree, etc. — and gently take up the slack. Tension the stretcher until the wire is tight, then staple it securely to the corner post. Staple wire to line posts, but leave staples slightly proud so that the wire can float; this insures that if an animal impacts the wire, it will be less likely to break.

Continue around the length of your entire fence, and there you have it — the really hard work is completed.

Gating Your Community

Most folks choose to hang ready-made pipe gates on their fences even when they use homestead-harvested posts. You'll certainly save some time doing it that way, but time can come with a fairly hefty cost. If you need a gate right away, by all means buy one — or use a carefully wired piece of welded stock panel until you get around to making one. In any case, making a gate is not only fun, useful, and visually rewarding — but you can also adapt the techniques you'll use to creating hurdles (an old-fashioned word for stock panels) of sufficient strength to temporarily enclose your flock of sheep and goats or a few head of gentle cattle. You can also create sorting panels for moving hogs around — and so much more.

Our approach to building gates is to select suitable rot-resistant saplings up to about 3 inches in diameter for the standards (vertical members), bottom rails, and diagonals. The same material can be used for the remaining rails, but you can also use lighter-weight, less rot-resistant species for those less-critical components that won't be in contact with the ground. Three-inch-diameter saplings should be harvested green and carefully split in half lengthwise before they have a chance to dry out and get hard. Attaching rails to standards and diagonals can be done with wire, line, nails, screws, or by cutting mortise-and-tenon joints and pinning them. Use what you have, and let your skill set guide you. For a couple of specific examples, check out the sidebars, "Homemade Pasture Gate" and "Rustic People Gate" in this chapter.

In the case of hurdles or lightweight stock panels, you can create what amounts to gates with elongated standards with both ends pointed at the bottoms. If you make them 8–10 feet in length, five or six of them wired together will contain a small sheep flock quite nicely (see Figure 6.9).

Homemade Pasture Gate

Our sheep have a stubborn streak that makes it tough to move them to greener pastures without a fight, especially when the pasture gates aren't tough and tight. Recently, we had the bright idea to move the flock to the east pasture for a spell of late-season grazing. Less than two hours into the experiment, half the flock discovered that the ancient barbed wire-and-batten gate separating our backyard from their pasture was loose enough to squirm right under. When challenged, the girls slipped back into their pasture without incident, and a judiciously placed welded-wire stock panel kept the lambs off the lam for the immediate future, but we really needed a better solution.

Wincing at the notion of spending more than 120 hard-earned bucks for a 14-foot pipe gate, we decided to build a rustic version instead.

The following morning, after a bit of measuring and figuring, we headed off to the woodlot and cut sufficient Osage orange and hackberry saplings to make a pair of five-bar pasture gates that would collectively span 14 feet, meeting in the middle. We chose decay-resistant Osage orange for the gate's vertical standards and the top and bottom rails; hackberry made do for the inside rails because it is lighter (and we had more of it).

The first step was to cut a pair of standards for each gate — ours are about 5½ feet long on the hinge edge and 4½ feet long on the latch edge. Next, we trimmed the top and bottom rails to length, shaped their ends with a hatchet, and nailed them to the standards with 16-penny nails. Once assembled, we racked the gate frames until diagonal measurements were within ¼ inch of one another and called them square.

Next, we peeled the bark from several hackberry poles using a drawknife (bark-on hackberry poles rot quickly), then shaped and nailed them to ☞

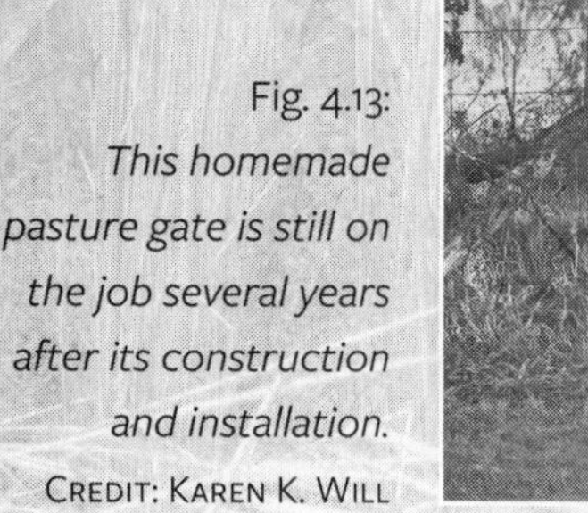

Fig. 4.13: *This homemade pasture gate is still on the job several years after its construction and installation.* Credit: Karen K. Will

the squared frames. With two hackberry rails installed, we noted that the gates were getting heavy, so we substituted twisted barbed wire in place of the planned fifth rail, which keeps the cattle from poking their heads through the gap.

We hung the gates using a time-honored method of planting the long end of the hinge standard in a hole next to the fence post to serve as the bottom hinge and wrapping a double loop of smooth wire around the post and standard for the top hinge. Here in Kansas, Osage orange poles stuck in the ground should last for about 25 years before rotting away, at which point we can install proper forged hinges if we wish.

Shortly after hanging the second gate, the sheep came bleating up to visit. I don't know whether it was the heavy-duty look of the gates or that the spaces between the rails were sufficiently narrow, but the entire flock went back to grazing without even giving our homemade pasture gate a test.

Rustic People Gate

A rustic gate made with materials harvested from your woodlot is a perfect finishing touch to your rustic fence. Structures like gates can make use of tall, thin saplings or other trees that might otherwise just seem a nuisance.

The first step is to walk your wooded areas or fencerows in search of materials. Long-lasting gates can be constructed with almost any tree species since they will not generally be in contact with the ground. Species like Osage orange and white oak are both strong and decay resistant, and they are relatively easy to work with hand tools when green.

For the gate's uprights (standards) select saplings that are 3 inches in diameter and cut standards about as long as your fence is high. Before the billets have time to dry out, carefully split them ☞

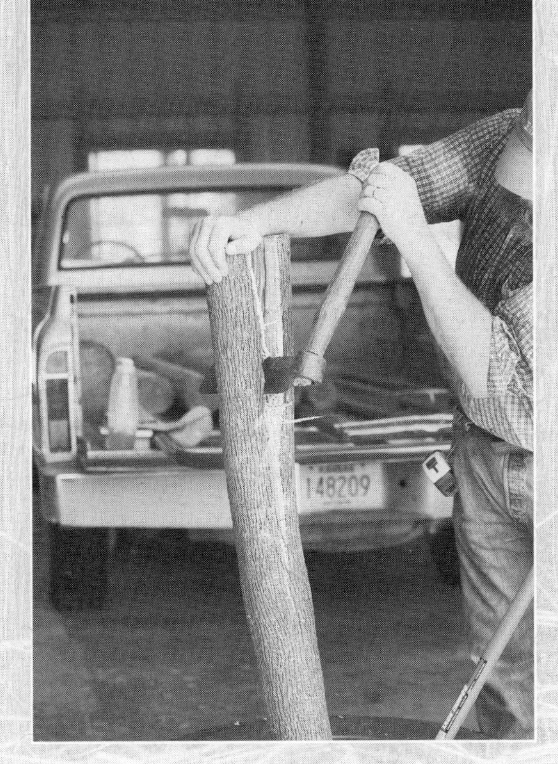

Fig. 4.14: *Using a froe to rive out a pair of Osage orange gate standards.* Credit: Karen K. Will

Fig. 4.15: *Pinned mortise-and-tenon detail on a homemade people gate. Similar methods would be used to assemble hurdles.* CREDIT: KAREN K. WILL

down the middle (from top to bottom) using a splitting maul, froe, wedges or better yet, a combination of all three. If there is any curve to the billets, you can make creative use of symmetry if you split the standards along an axis perpendicular to the curve.

The gate's rails (horizontal), pickets (vertical), and diagonal brace can be harvested from split saplings about 2 inches in diameter or split from larger diameter materials. If the saplings are of a fairly decay-prone species like hackberry, peel the bark from the gate pieces to slow the process.

Working on the split faces of the gate standards, mark mortises about ¾ inch wide by 1½ inches long about 3 inches from the ends. Using a drill and chisels, chop through-mortises — a slight narrowing taper from marked side to opposite side will make crudely chopped tapered tenons fit tightly.

Using a hatchet, cooper's axe, one-handed adze, drawknife, or other handy tool, pare down the ends of your rails to form tapered tenons about ¾ inch by 1½ inches in cross section at their inboard ends. Fit these tenons into the through-mortises you cut into the standards in the step above. For best results, you want the rails and standards to be approximately perpendicular. Pin the rails into place with nails (be sure to first bore pilot holes for the nails or risk splitting the relatively thin wood).

Lay this rectangular frame on a level surface and square it by measuring its diagonal dimensions. When the diagonal measurements are identical or close to it, the frame is square enough. To keep it that way, cut a diagonal brace to fit tightly in the crooks formed by the intersection of the standard and rail on the bottom of the hinge side and the top of the latch side of the gate. Fasten this brace in place using nails and/or wire. Using a branch with a small Y at one end and whittling a V at the other end can really strengthen your gate's diagonal brace.

Split sufficient pickets to make your gate as tight as you need it to be. Arrange the pickets symmetrically to take advantage of any curves or tapers in the pickets, and nail them to the standards and diagonal brace; again, be sure to bore pilot holes if there is any risk of unwanted splitting.

Finally, install the hinges, hang your gate, and enjoy its functional beauty for years to come.

Build a Garden Fence for Free

Our homesteading friends, Eric and Wendy Slatt, are no strangers to making do around their place, so when it came time to fence the critters out of the kitchen garden, they took a good look at their wooded acreage for ideas. In the end, the Slatts decided to create a picket-type fence using rot-resistant species for post material, seedlings of lighter scantling for the two rails, and even lighter material to make the pickets. Almost no metal fasteners were used for this project — Eric lashed the rails to the posts and the pickets to the rails first with baler twine and later with nylon mason's line once the twine began to rot away. Later, they installed welded wire to the inside of the pickets to help with rabbit control. The project took some time, but the end result is as lovely as it is functional.

Fig. 4.16: *Lovely rustic picket-style fence even keeps out the rabbits when reinforced with wire mesh along the bottom.*
CREDIT: WENDY SLATT

Easy Homemade Stacked Rail Gate

This gate takes its building cue from the old stacked rail style split-rail fence and is quite secure for most livestock, so long as they don't get the notion to rub on the ends of the gate rails and knock them off the posts. Since this method borrows from the stacked rail fence design to close an opening for a wire fence, you'll need to double up on the posts that define the opening. First, set and brace the anchor posts in the fence line that will define the opening. Next set two additional posts offset ☞

about a foot perpendicular to first pair of posts, relative to the fence line. These posts may reside inside or outside the fenced area depending on which side of the anchor posts you set them. Next, lash, bolt, nail, or otherwise attach short horizontal support timbers running from the anchor post to the adjacent offset post with their top edges at the heights that you want your gate's rails to be located. Finally cut rails from the woodlot of sufficient length to insert into the spaces between adjacent posts and rest them on the support timbers. This gate takes some time to open and close, but for seldom-used openings, or when money and/or hardware are in very short supply, this is an effective gate that requires little more than some energy and time.

Fig. 4.17: *Stacked rail gate, closed.*
CREDIT: WENDY SLATT

Fig. 4.18: *Stacked rail gate, open.*
CREDIT: WENDY SLATT

Chapter 5

Building What You Need

In the last chapter, we considered using the woodlot as a resource for creating sturdy fencing, but with a few hand and/or power tools, you can also create useful animal shelters and other objects with relative ease and little investment. Need a new lean-to or other small pole building? How about that workbench project? Kitchen island? Wherever trees grow, you can harvest them and use them directly as building timbers for small structures or saw them into lumber for use in all kinds of construction projects. And you don't need to be a fine carpenter to make it all happen. Armed with the knowledge presented here, you'll no longer look at a snag (a dead, but still-standing tree) as just so-much firewood — there's much more to it than that. As your woodworking ardor develops, please read all you can, take a workshop or three, and don't be afraid to jump in and try something new.

Tools of the Trade

Check any periodical on fine woodworking, timber framing, or wood shaping, and you'll notice that even the oldest crafts now make use of more modern tools; indeed, some of the crafts are characterized with a steady onslaught of new tools designed to make old techniques easier to master for the average weekend warrior. But don't be discouraged if you don't have a power mortiser or pocket screw jig when those tools are recommended for a particular project. All you really need to create beautifully functional structures from your woodlot is a set of relatively simple hand tools. You can

Fig. 5.1: *Drawknifes, a heavy mallet, adzes, augers, hand drills, pinch dogs — you name it — all have a place in the homestead's hand-tool chest.* CREDIT: OSCAR H. WILL III

Fig. 5.2: *In addition to heavier saws used in the woodlot and working with heavy lumber, the small dovetail saw, keyhole saw, and coping saw (missing its blade) will handle much of the finer wood cutting work.* CREDIT: KAREN K. WILL

add various power tools as opportunity and real need arise. For example, you can cut fence posts or even lumber to frame small buildings using any number of hand saws. You'll get a workout using a saw, but you'll also be saving money on your gym membership (see Section 3).

Cutting Tools: Must-have tools for the average homesteader include a saw for cutting trees down and into lengths. If the stock you have to deal with tends to be in the 6–8 inch diameter range, you might get away with an inexpensive bow saw. By lurking at farm sales and junk shops, you might also net a one- or two-person crosscut saw for short money. You'll need to learn about setting and filing hand saw teeth if you go the antique route, but those skills will keep you going long after conventional energy becomes excessively costly — or even unavailable. If you have lots of rough cutting to do, investing in a chainsaw probably makes sense, if for no other reason than the time it can save. If you go with a chainsaw, be sure to invest in extra chains and the sharpening tools you'll need to keep them in good shape.

Bow saws and chainsaws will get your wood out of the woods in good shape, but if you want to make finer cuts for building things like hay rakes or kitchen islands, be sure to invest in at least two hand saws designed for cutting lumber and smaller timbers.

One should work well across the grain (a *crosscut saw*) and the other along the grain (a *ripping saw*). Later on, you might decide to go higher-tech. We find our 10-inch table saw to be indispensible for furniture making, but there was a time when the only power saw we had access to was a handheld circular saw with a combination crosscut/ripping blade. If you have only hand saws, you can still build — it will simply take a bit longer.

Other cutting tools you might consider acquiring include a *power chop saw* for easily cutting steel and other metals. You can use a hand *hacksaw* to do that work — with a few more calories burned and some time. If you plan to cut lots of intricate shapes in relatively thin wood, look for a small *power band saw, handheld power jig* or *saber saw,* or the versatile and inexpensive *coping saw.* You can also fit thin blades into some antique wooden bow saws that allow you to cut sufficient curves to make pleasing chair-back slats.

Hatchets and axes can certainly fall into the cutting-tool category, but we find these devices to be more useful in the wood-shaping department. If a felling axe is what you have, by all means put it to use for harvesting wood and cutting logs or timbers to length.

De-barking tools: Since the wood of most tree species lasts much longer if the bark is removed, consider keeping a *barking spud* in your toolkit along with a robust *drawknife* that can be dedicated to barking (although it can serve double-duty as a shaping tool in a pinch). If you can't afford or find a barking spud, or the right drawknife, you can alter the rear axle shaft from an old three-quarter ton or one-ton capacity pickup truck (if it's a model with a full-floating rear axle). Heat one end to beyond red hot in a good campfire and hammer it into a rounded, flat chisel shape (or have your friend at the local welding or blacksmith shop do it

Fig. 5.3: *The large drawknife on the left is useful as a barking tool and for hogging lots of material when shaping large timbers.*
CREDIT: OSCAR H. WILL III

for you). Let the piece cool slowly, and it will be soft and relatively easy to shape with a file or grinder. The piece will be heavier than wooden-handled models, but it will make bark removal go fairly quickly.

A de-barking drawknife can also be fashioned from a length of relatively light automotive leaf spring. Don't worry about the curve. Take a single leaf from the spring and cut a 12- to 18-inch-long section out of it. Using a grinder, grindstone, and files, remove metal from one side of one of the long edges of the spring. You want something that looks like a one-sided knife edge in cross section. If you only want to use this as a barking tool, don't worry about being too particular with the final edge — as long as you can get it sharp and keep it that way with a file. Next, select two lengths of steel rod from your scrap pile and weld them to the ends of your sharpened spring to serve as handles. The actual length can be determined based on your hand size. If you don't weld, your local welding shop (or neighbor) can probably do it for you for the minimum shop charge, a few pounds of barter goods, or a barter service. Fit wood, bicycle grips, or some other forgiving material over the steel handles, and you will have yourself a mighty functional de-barking drawknife.

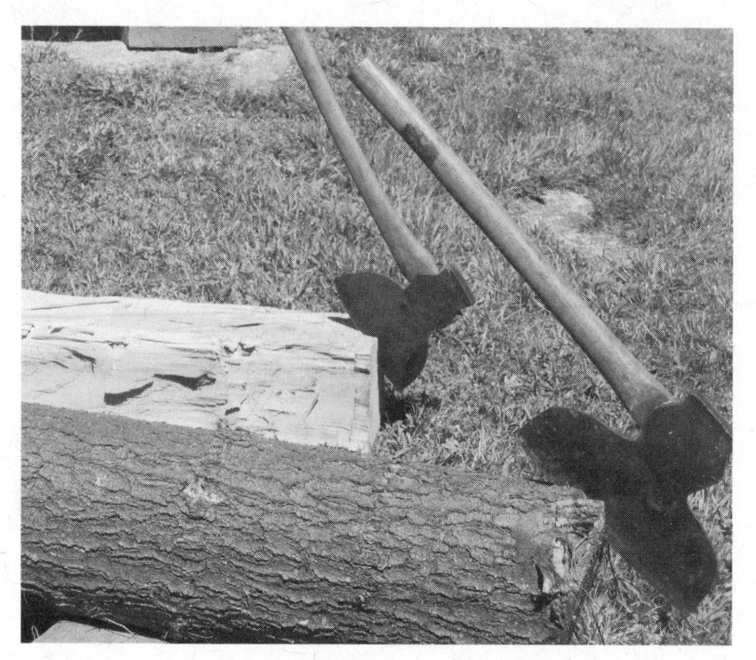

Fig. 5.4: *The single-handed broad axe ("broad hatchet," in some references) and two-handed broad axe (right) are both useful for shaping timbers.* CREDIT: OSCAR H. WILL III

We've left out all the metallurgy here, so as not to discourage you, but if you are interested in changing the temper and hardness characteristics of either homemade version of the debarking tool, read all you can about the colors associated with temper, and then experiment with heating, quenching, and tempering steel.

Shaping tools: Whether you're building a gate, bench, or barn, you'll need to do some shaping of wood, and this is a place where a small set of good hand tools will really shine. And yes, there are some powered alternatives, but in our opinion they don't offer that much in the way of time-savings, convenience, or increased pleasure in the work.

The homestead woodshop should definitely contain a couple of sets of robust chisels. For really hard chopping and for digging out mortises, a set of four or more *mortising chisels* ranging in width from ¼-inch to 1-inch should cover most of your needs. Mortising chisels are characterized by having full-width, non-tapered (or very slightly tapered) sides. They tend to be at least ¼-inch thick, so they won't bend when you're digging out mortises in a prying fashion. Mortising chisels are available from specialty hand-tool vendors, but you might also find some nice ones at flea markets or antique shops.

The second set worth collecting sooner rather than later would be a set of *bevel-edged bench chisels*. These are available practically anywhere hand tools are sold. If you purchase them new, err on the side of robust construction and handles that can take some striking without breaking to bits.

Armed with the chisels just described, you'll be able to chop mortises, pare tenons, clean up all kinds of joints, round corners, shave little bits of material here and there, and so much more. Resist the temptation to purchase a mallet to use with these chisels — you can readily make one from your woodlot.

When creating wooden bowls or other hollowed-out objects, a set of *gouges* can be very handy. Gouges are essentially curved chisels that you can use by hand alone or with the help of a mallet to take a split log and shape it into a dough bowl. It's definitely worth having a small collection of gouges on

Fig. 5.5: *Mortising chisels are considerably heavier in cross section than firmer and bench chisels, which keeps them from bending or breaking when prying chips from the mortise. Note the diamond-coated sharpening stone propping up the chisels.*
Credit: Karen K. Will

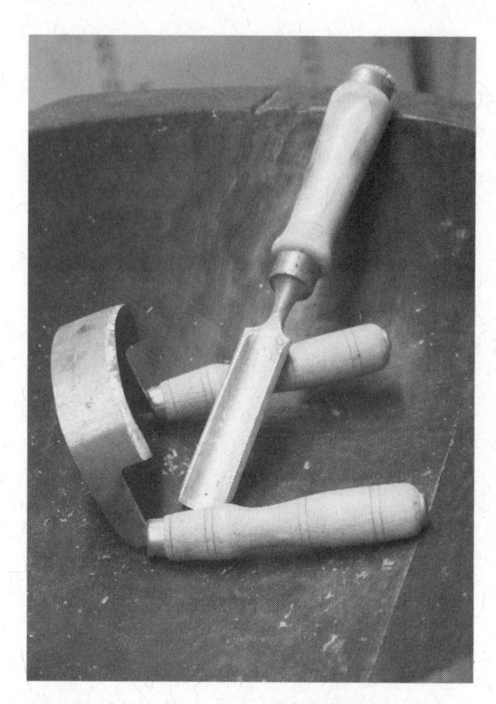

Fig. 5.6: *If you plan to do much bowl or dough tray shaping, the inshave and gouge (curved chisel) will come in very handy.* Credit: Karen K. Will

Fig. 5.7: *If you encounter a forge-welded froe and an old cooper's axe at an antique store or farm sale and they aren't too pitted, by all means snap them up, clean them up and put them back into service. Although much newer, the rusty sculptor's axe comes in very handy when shaping and hollowing smaller projects.* Credit: Karen K. Will

hand, but you probably won't reach for them as often as you do a chisel.

If you plan to work with green wood, a *froe* and a drawknife are must-haves. The froe is an old-fashioned, flat-bladed (double-bevel edge) tool with a vertical handle that's used to split billets and rive out more dimensionally appropriate pieces for various projects. Shingle makers used the froe to split wooden shingles; some chairmakers use the froe to split out blanks for chair backs, legs, rungs, etc. The froe will take the place of a saw when splitting a blank into slats or rungs — and, since the wood splits along the grain, the resulting pieces tend to be stronger than sawn pieces of similar scantling. The froe is handy for making gates and hurdles, hay rakes, stools — you name it.

Some relatively light-duty welded froes are available at specialty woodworking supply houses, but try looking for a forged antique one (if you're lucky, it will have been hand forged). You can tell it was forged if the handle loop is formed from the same piece of metal as the blade and is turned back onto the blade with the end hammer-welded to it. You can also make a froe from scraps — a straight piece of automotive leaf spring heated, shaped, and ground will do the trick. Or you can take any heavy bar stock (¼ inch thick or so), shape the blade with a grinder, and weld a short length of heavy-walled round tube to the end to form the loop for the handle. And for the handle, just take a look in your woodlot. You're sure to find something that's not just suitable, but lovely to boot — and free, for a bit of labor.

Sometimes you'll have the need to split a longer billet, and the froe might need some help. Keep that in mind when you are at junk shops and farm sales, and build your collection of *splitting wedges*. The wedges are useful for splitting firewood, riving, and directional felling. Even if you have only a pair of steel wedges, you can supplement your wedge supply with handmade wooden wedges called *gluts*. You can't use a glut to open a split very effectively, but you can use it to hold a split open to make easier work with the froe or for repositioning your steel wedges. Cut any number of gluts from hardwood limbs — smooth the wedge edges so that they will be less prone to catching and getting destroyed while driving them.

It's difficult to imagine working with raw materials more or less direct from the hedgerow without a drawknife and at least one *spokeshave*. A drawknife is a heavy single-bevel knife (straight or gently curved) with a handle at each end. It's used to remove shavings from wood with virtually any cross-sectional shape, most often by pulling the tool toward you (but don't let that description keep you from pushing the tool when and where it is effective). The drawknife is generally designed to be drawn with the bevel side of the blade up, but there are plenty of occasions to use it with the bevel side down. When the bevel is up, it will take more aggressive cuts.

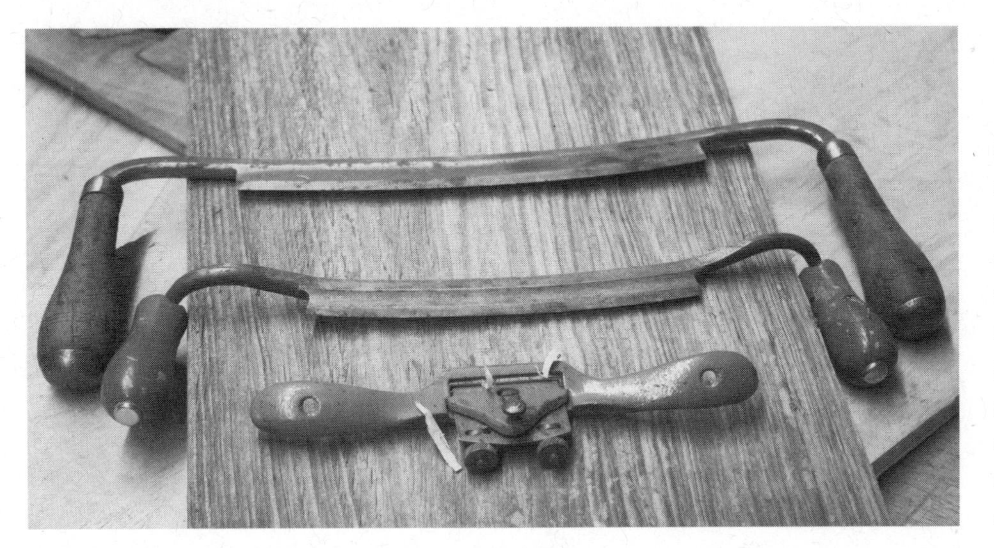

Fig. 5.8: *The flat-soled spokeshave (bottom) is particularly suited to rounding spindles, tool handles and wagon wheel spokes but it can also be used to smooth curves and can be used in conjunction with finer drawknives.*

CREDIT: KAREN K. WILL

The drawknife is used around the homestead to de-bark saplings, shape tool handles, chair rungs, bench legs, and the like. It is also quite useful for cleaning up any rough edges when you are finished working with the froe and for changing square or rectangular cross sections of wood into oval or round cross sections. The drawknife can also be used to carefully shape table or bench edges, the bottoms of wooden bowls or platters, and to create square or rectangular cross-sectioned tenons on rounded stock. Although it's often missing from the modern toolbox, it's virtually impossible for us to imagine life on the modern homestead without one or two drawknives to call on.

When your work requires creating lots of rounded spoke-like objects, the *spokeshave* is a must. We keep a flat-soled (flat bottom) spokeshave in our shop, although round-soled and combination-soled versions are available. The spokeshave is essentially a very short-soled plane with handles attached to both sides — it can be pulled or pushed through simple or compound curves. If you are making Shaker-style chairs, you'll likely clean up the drawknife work with spokeshaves when finishing legs and rungs. You might also use a spokeshave to make really nice hay-rake teeth or dowels that you can use for any number of purposes.

The *inshave* is a wonderful tool that you might add to your toolbox if you start making a cottage business of carving dough bowls or plates. This tool is much like a drawknife except its blade is shaped like a "U." If you get into hand-hollowing in a big way, you might add a *scorp* or two to your kit — these tools are like the inshave

Fig. 5.9: *Cast iron or steel planes come in all sizes and quality levels. This collection varies in length from the longest, a jointer (or try) plane to the smallest, a block plane.*
CREDIT: KAREN K. WILL

except they are operated with a single hand (generally), and take smaller but potentially deeper and steeper bites.

Entire books have been written on *hand planes,* their proper use, tuning them up to get precision cuts, etc. For the homesteader, this is all interesting and useful information, but essentially, you want to add a plane or three to your toolkit to help keep things smooth, to make straight edges, and to adjust the thickness of your work pieces. Your planes will be reserved for finish work or work of a finer nature. You might use a *jointer plane* (plane with a sole approaching 22 inches long) to level your workbench's top or make the edges of boards you intend to join edge-to-edge perfectly straight. A *jack plane* (multipurpose plane with a sole in the 15-inch-long range) is useful in sizing, smoothing, and edging stock, and a *block plane* (small hand plane) is used for trimming end grain and fine sizing work. There are literally scores of different types of hand planes that run the gamut of smoothing and shaping work. Most consist of a sharp single-bevel blade held firm in an iron, steel, wooden, or metal/wood body that you move along or across the work, removing thin shavings as you go. As your skills and needs develop, you can add planes to your collection that will cut grooves, create tongues and shape decorative beading or molding. There are many more types of planes. You'll probably start to covet them if you get further into woodworking.

The venerable axe is still a very useful tool in the woodlot and wood shop. Felling axes can be found used quite easily; be sure to examine the head closely to make sure the bit (cutting end) and *poll* (opposite the cutting end) are not cracked or corroded beyond recognition. If you wind up with an antique axe head, it's the perfect time to collect your other cutting and shaping tools and make it a new handle. The felling axe has a double-bevel bit and is useful for dropping and limbing trees (a *limbing axe* is better for limbing, but plenty of trees get limbed with a felling axe) and for some rough shaping. Double-beveled cutting edges aren't

New Life Through Old Tools

With so many old timers and old farms ceasing to exist these days, old tools tell their life stories with poignancy. When we come across an old tool box filled with woodworking tools, or a collection of rusty, worn farm tools at an antique store or farm auction, we're compelled to save them from the landfill and give them new life with new work. Even if an old tool is well beyond repair, we've been known to buy it and use it as art somewhere around the farm. An old rusty tool with a handmade handle tells the story of a life and a place in every nick and pit, and one can only imagine all the projects that were completed, all the chores that were accomplished, and all the people that were satisfied by that simple, useful piece of iron.

always the best for shaping logs into timbers or cutting beveled tenons on the ends of gate rails.

If you could have only one axe in addition to that designed for felling, choose the *cooper's axe*. This broad-headed axe has a single-bevel bit and a subtle curve to the head such that you can engage a billet with the bit end while clearing your hand — and you can take shaving cuts because you don't have to work the axe at so sharp an angle in order for it to engage the work. The cooper's axe is really a one-handed broad axe; most of the old-fashioned types are symmetrical, so you can use them left or right handed depending on which end you install the handle.

There are many other axe and hatchet patterns out there. If you encounter them used, at a good price, by all means invest. If nothing else, you can oil them up, dress the cutting edge, install a homemade handle and sell them at a profit — or give them to your son or daughter as he or she puts together a homestead shop.

The *adze* is another shaping tool worthy of display on your woodshop wall. The two-handed adze looks a little like a mattock, except the end of the sweeping curve is sharpened like a chisel. The single-handed adze looks like a hatchet-sized version of the other. Some adzes have a curved blade that resembles a gouge — these are particularly useful in hollowing efforts. Adzes make short work of dressing/hewing beams directly from logs, hollowing out logs, shaping ends, cutting notches, and the like. You can even use the adze, along with a very large chisel called a *slick* to take squared-off timbers and make rounds of them. The adze is as comfortable in the sculptors' studio as in the wooden boat building or homestead shop.

Boring tools: Every homestead shop should have at least one *bit brace*, preferably with a universal chuck so that it can take the old-style boring bits with a tapered drive end as well as more modern bits with round or hexagonal ends. The bit brace coupled with a set of *ship auger bits* ranging in size from an eighth of an inch to at least an inch should serve most of your boring

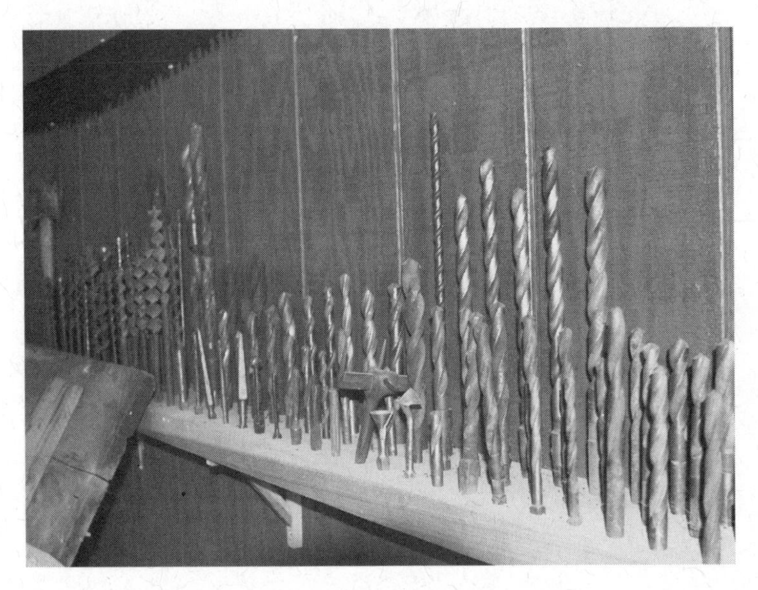

Fig. 5.10: *This collection of augers, twist bits and other cutting tools can be driven with bit brace, hand drill, drill press, and even a handheld power drill.* CREDIT: OSCAR H. WILL III

Fig. 5.11: *A pair of braces — note the round shank, which requires a combination chuck and the square shank (lower left), which requires a combination or traditional chuck.* CREDIT: KAREN K. WILL

needs. The ship auger bits are easy to keep sharp, and the screw tip draws them nicely into the work. For smaller holes, you might find an old hand drill (think egg-beater-type contraption) and a set of high-speed steel *twist bits*. If you find yourself boring lots of holes and have the budget, a corded or cordless hand drill can shave hours of time off more elaborate projects — and there are many more bit options (some are disposable) available for use with power drills. Still, we reach for the bit and brace when building fences far away from the power lines.

Sharpening tools: Though often forgotten (or, at the very least, neglected) in this throwaway world of ours, keeping your hand tools in tip-top cutting, shaping, and boring condition will require a number of sharpening devices.

A collection of medium and fine (as opposed to coarse) metal files definitely has its place in the shop. Flat files can be used to dress axe bits and other relatively long-curved cutting edges such as the drawknife. Very narrow triangular cross-section files are great for keeping your ship auger bits sharp. Round cross-section files can also be used to dress chainsaw chains, although it's wise to use a dedicated grinder for that work since the chainsaw is used so regularly.

In addition to files, conically-shaped stones (or diamond-coated metal shapes) called *slips* should be on hand to keep tools such as gouges nice and sharp.

Likewise, if you have any specialty carving tools or chisels, a slip stone shaped to match the blade's cutting edge is invaluable. Flat stones or slivers of flat stones can be used to dress these tools, but it takes great patience and skill to do it well.

Knives, plane irons, straight or skewed chisels, and most other cutting edges, including those found on high-speed twist drill bits, can be kept sharp with a set of flat stones of medium and fine grits. The stones can be natural, synthetic, diamond-coated metal, or diamond-impregnated polymer. Some of these sharpening devices work best when soaked in water first or with their surface kept wet with water or cutting oil.

If you cannot afford a decent set of sharpening stones, you can substitute a piece of machinist's granite (ground very flat) on which to adhere finer grits of sandpaper (designed for wet sanding). Cut the sheet of sandpaper to size, spritz some water on the polished surface, and press the sandpaper's back to the stone. If you can't find a piece of granite, you can use heavy plate glass — so-called *float glass* is ideal, but if that's not available, use the heaviest, flattest piece of glass you can find. Once the abrasive paper is adhered to the flat stone or glass, simply use it like you would a whetstone.

When it comes to power bench or stand grinders, begin with a slow-speed, wet stone if you can. These grinders turn slowly, and the wheels are bathed in cooling water, which keeps your tools from heating up and losing temper. We generally only use a power grinder when we inadvertently nick a cutting edge, and we need to remove relatively large amounts of material more quickly than is possible with hand stones and files. If you only have access to a high-speed bench grinder, by all means put it to use, but use it very carefully. Dip the tool that's getting sharpened into cooling water regularly; you don't want to cause the cutting edge to turn blue from heat.

Look in any tool catalog today and you'll find literally thousands of dollars worth of different devices designed to help you keep your edges sharp. Some are magnificent; others are amazing. But if time is your only savings, why not spend a lovely evening in your shop — woodstove turned down low, your favorite music playing — and really get to know your cutting, boring, and shaping tools by maintaining them with your own hands?

Grow Your Own Lumber

We're fortunate to have sufficient trees growing on our Kansas farm to provide fence posts, pole-building posts, firewood, and saplings for building gates and related structures. We also have many dead or dying hard pines and a couple of hardwood tree snags where the bulldozer piled debris after repairing a couple of pond dams. And, as much fun as it is to work with green wood and hand tools, there are times when what you really need is dimensional lumber — or at least lumber in the specific dimensions you desire. Take that one step further, and, if you have sufficient excess trees, converting parts of them into lumber can help bring a little cash to the homestead. And we all need some cash.

While pondering a home-improvement project one day, an urgent need for lumber led us to investigate small portable sawmills rather than pay the prices for lumber at the local home improvement store. In fairly short order, we had a small Granberg Alaskan chainsaw mill in hand, complete with a

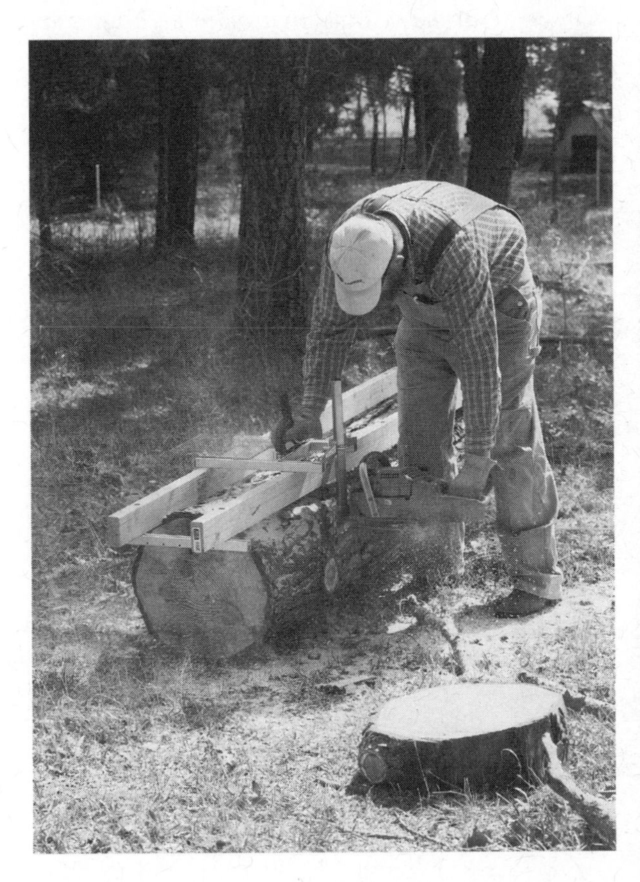

Fig. 5.12: *The Alaskan chainsaw mill in action. It took about two minutes to make 10-foot-long cuts in this hard pine log.* Credit: Karen K. Will

Fig. 5.13: *Turning logs into timbers is easy and fun with the Alaskan chainsaw mill.* Credit: Karen K. Will

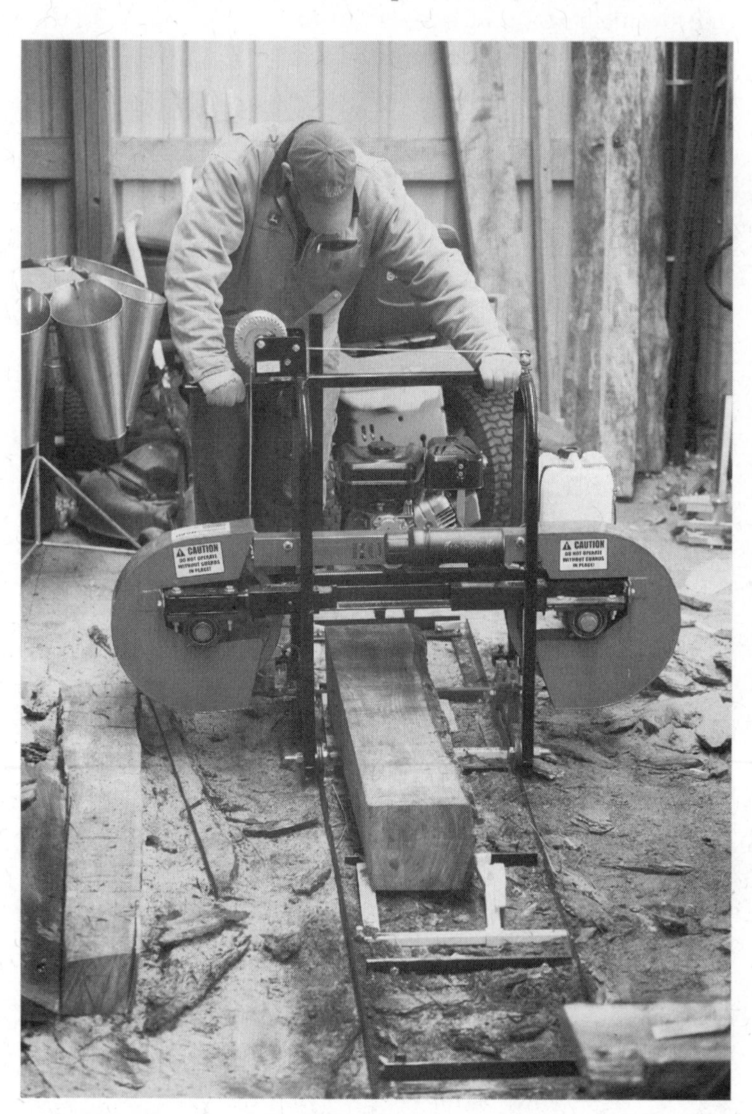

Fig. 5.14: *The Hud-Son bandsaw mill made short the work of sawing black walnut for the kitchen island's top.*
CREDIT: KAREN K. WILL

pair of ripping chains, ripping bar, and instructions on how to mount it on our pro-model Husqvarna chainsaw. The entire package cost less than $200 (in 2010). With about one hour's worth of cutting, we had a large accumulation of hard pine timbers and boards — enough to build the frame, doors, drawers, and legs of a kitchen island. The pine was milled from the butt end of a tree that had been dead for three years. We stacked the resulting lumber in our barn, where it sat for about seven months and reached a moisture content of about 9 percent. With the help of a table saw, a power planer (bought at a yard sale), circular saw, and a mess of hand tools, that material eventually became the base cabinet for the island. We also milled some American black walnut (from snags) for the project's top.

In the meantime, the Hud-Son tool company sent me a homestead-model bandsaw mill to try out as part of my job at *GRIT* magazine. This machine is lightweight, easy to set up, easy to use, and capable of handling 22-inch-diameter logs. The bandsaw mill is a little faster and wastes less wood in the form of sawdust, but it costs about ten times the price of the Alaskan mill. Finishing boards using hand planes is also easier with the bandsaw mill because the tooth scars are not as deep. Looking back, we were glad to have had both machines. If we planned to mill lots of wood into lumber, it would be tough not to choose the bandsaw mill, but again, time and a bit more wastage is a small price to pay for a 10-fold initial payout.

It would not be difficult to invest in an Alaskan mill, use it for a number of projects and/or make lumber to sell in order to pay for the purchase a small bandsaw mill. That makes a lot more sense than purchasing the expensive mill on credit, unless you have a line of paying customers already in place.

From Woodlot to Kitchen Island

Access to a sawmill makes it possible to create dimensioned lumber for all kinds of projects, including basic cabinet making. In our case, we also had a thickness planer that could be used to do some initial sizing and smoothing of the boards as needed. Building with homemade lumber takes a little more time, but the payoff is huge in satisfaction, price, and the fact that you get to control the dimensions of the boards. Our first woodlot-to-lumber project was motivated by the need for additional storage and surface space in the kitchen — plus the desire for something truly unique and evocative of our land.

The first step in the process was to generate a rough design. We decided on a 34-inch-wide by 42-inch-long base footprint that would be 2½ inches shorter than the other countertops in the kitchen. The top would be 2½-inch-thick solid walnut, 46 inches wide and 44 inches long. The extra width was for an overhang where we planned to place a pair of stools. The base and top put together would be the height of the other countertops in the kitchen.

The project began with sawing heavy 3¼-inch-square leg material and 1¾-inch-thick rail material; the table saw was invaluable for this work. Next, I ran the pieces through the thickness planer to bring the dimensions down to 3-inch-square for the legs and 1½ inches thick for the rails. Hand planes would work fine for this cleanup and sizing, but our planer shaved quite a bit of time off the process.

The table saw was again put to use cutting the tenons for the rails. For this cabinet's framing, I used through-tenons let into the legs. I cut the mortises using a ¾-inch bit chucked into the drill press (another yard-sale find) and cleaned them out with an old set of mortising chisels found at a flea market about 30 years ago (you never know when things will come in handy). The mortises could easily have been cut with a ship auger and bit brace or even a handheld power drill, but the drill press makes it easy to keep the bit at a perfect right angle to the work.

Once the base was framed up, I installed ½-inch-thick planks for the sides and back and a somewhat thicker bottom. Then we built the doors and drawers, and then we moved on to the top.

For the top, we chose our best pieces of 2¾-inch-thick black walnut. We weren't too happy when, just as I was running them through the thickness planer, it decided to eat part of its armature, which rendered it completely inoperable. Repair parts were still available, but they would set us back a couple hundred dollars, so I decided it was time to tune up the hand planes. Since we don't own a power jointer, I needed the hand planes anyway to create smooth, flat mating edges between the walnut planks, so they could be glued together.

Once I had the walnut prepped, I glued the pieces edgewise with epoxy. Once the glue cured, I trimmed the top to size, cut rebates on each end to form a tongue that I fit into a groove I cut into "breadboard" ends. (I went with breadboard ends to minimize the amount of end grain that the island top would have exposed, and to help keep those heavy planks aligned as they adjusted to seasonal moisture changes.)

With the top assembled, I spent a few hours working it flat with the planes, sanded it smooth, and encapsulated the entire thing with epoxy. The epoxy was really overkill, but it will help keep the thick top dimensionally stable. More sanding and several coats of polyurethane later, and we had a lovely, kitchen-friendly island top.

The final stages of the project included finishing the base. Karen first stained the raw pine with a dark walnut color. After that dried for two weeks, she lightly sanded and then painted the entire base with a white, pearl-finish latex paint. Once the paint fully cured, to give the island an aged, rustic appearance, she brushed on a dark walnut wiping stain — working in sections, followed by immediately rubbing it off.

Fig. 5.15: *Completed kitchen island.* CREDIT: KAREN K. WILL

Used in this manner, the wiping stain acts as a glaze, and it settles beautifully in all the handmade gouges and planer marks. We ordered some hand-forged towel bars and hooks from a blacksmith (on Etsy) to install on both ends. It took most of a year here and there, but we have a kitchen island like no other, and our only costs were in epoxy, finishes, hardware, and a bit of fuel.

From Woodlot to Loafing Shed

The way our farm was configured a few years back, we really needed one more livestock shed for lambing and to keep the guard donkeys out of the rain, so we decided to build a *loafing shed* using found materials. The starting point for the loafing shed was an old native-limestone hog shed foundation and walls. The old hog shed itself was long gone, but the walls were still sound, and the 100-year-old anchor bolts were also still intact. We didn't want to put a roof over the entire 32 feet of shed, so we decided to just cover half of it. The covered area is about 16 feet by 16 feet.

The first step in building the loafing shed was to source the posts and timbers. Since we had just renovated a couple of pond dams, we had plenty of fairly straight hackberry trunks on hand, so we collected five that were roughly 6 inches in diameter and more than 16 feet long, and two that were closer to 8 inches in diameter and 18 feet long. Because of the dense thicket that these hackberries had been growing in, they were thin, fairly uniform

Fig. 5.16: *Loafing shed — note the roof had a close encounter with our loader tractor and so is missing a notch on the right side.* Credit: Karen K. Will

in diameter, straight, and tall. The five thinner hackberry trunks were designated roof purlins, and the two larger diameter pieces would become the rafters. Around here, hackberry left in debris piles tends to rot fairly quickly, so we checked them carefully and deemed them sound. It would have been prudent to strip the bark off those timbers, but we were in a hurry and did not. The bark has subsequently separated nicely; and, because they're out of the weather, sub-bark decay is virtually nonexistent.

For the posts, we headed to one of our deeper hedgerows and managed to find a single Osage orange tree that had a pair of long and fairly straight 12-inch (at the base) diameter sucker trunks growing from what amounts to a stump left when a much larger diameter single-trunk tree was harvested quite a few years ago. Osage orange is an excellent coppicing species, so this once-cut tree actually sprouted many new trunks that grew quite straight and relatively tall. I cut the two trunks and trimmed them to roughly 13 feet long. I had to use a tractor to drag these monsters back to the building site.

Osage orange is perfect for posts because it is so naturally rot-resistant and strong. Long before the days of creosote and other pressure treatment of hard and soft woods, Osage orange was prized for making railroad ties. Some lore suggests that the native species was almost driven to extinction by over-harvesting for that purpose. Luckily, many of those stumps sprouted and yielded a second crop of fine fence posts, just in time for barbed wire to show up on the Central Plains. In our part of the world, a 4-inch-diameter Osage orange fencepost, planted with its bark on, will last around 30 years. And, the thicker the post, the longer it takes to rot through.

Sinking those big posts to the first limestone layer, about 5 feet down, would have been much more difficult without the help of a loaner backhoe. What took roughly a half hour might have taken more like a day had the digging been by hand — but it would have been entirely doable to set the posts into hand-dug holes. (It would have been much easier had I been 30 years younger, too, no doubt.) We planted those posts roughly 12 feet apart at the front of the former shed — yes, the tractor loader was invaluable in getting the posts into the holes and holding them more or less plumb while I backfilled. It took much more time to backfill the holes because we wanted those posts to resist racking and wobbling without a bunch of diagonal bracing. Essentially, I shoveled a couple of inches of soil back into the hole and packed it with a long stick having a cross section about an inch square. By the time we were finished, the posts were rock solid. Interestingly, most of the soil that had come out of the hole went back into it. Talk about compaction.

The next task was to mark and notch the locations of the rafters on the outside edges of the posts. I cut the notches with a chainsaw and a single-handed adze and pared the sides of the rafters flat to match. The loader came in handy with setting those 18-foot-long hackberry rafters — I set one end on the back wall and raised the front ends until they mated with the notches on the posts. Next, I bored ¾-inch holes completely through the rafter and the post, drove a ¾-inch by 12-inch carriage bolt through the works and added a flat washer and nut. The rafter wall ends were secured with perforated strapping and the half-inch anchor bolts left in the old wall.

Once the rafters were set, we installed five 16-foot-long hackberry poles perpendicular to the rafters to serve as purlins. I did a little notching to help keep the pitch fairly smooth (to aid with roofing) and a little trimming of knobs and high-spots on those purlins for the same reason. The chainsaw, reciprocating saw and adze were used for that work. The purlins were attached to the rafters with ½-inch-diameter carriage bolts, washers, and nuts.

Even though we have a pile of not-too-badly rusted corrugated metal that came from another shed tear-down, we opted to go with new roofing material because of the wind — and because the roof is visible from the road. It took about $120 and an hour of time to grab the galvanized, corrugated steel at our local building supply. We used 16-foot lengths of the material to avoid making too many joints. The roofing was attached to the purlins using ring-shank nails with synthetic rubber gaskets under their heads (the kind designed specifically for nailing this kind of material).

Once installed, all the lumps and sways in the raw-timber roof framing were visible. I could have leveled things better and used some dimensional lumber to even things up a bit, but it looks whimsical. To date, the roof has withstood wind and snow, save for one small tear. Enclosing the front of the shed with panels and gates is still on the table, but in the meantime, we have a serviceable loafing/lambing shed that the donkeys seek out when the rain is cold, and the sheep pile under it when the sun is hot or when blizzards blow out of the north.

CHAPTER 6

You Can Build It

Our ancestors used their woodlots to provide material to make everything from oxen yokes to hay rakes to pitchforks — and you can, too. This chapter will consider the construction of hay rakes, benches, harvest boxes, pitchforks, and garden planting tools that can be readily constructed with woodlot materials at hand. Some projects are completely suited to using green materials more or less straight from the tree, while others might be easier or made better with lumber you saw from the trees. In any case, let these projects be a guide or provide stimulation for your own imagination, or serve as encouragement to just go for it.

Hay Rakes

If you're thinking about making hay by hand — scything, raking, and stacking — one of the tools you'll need is a light, functional hay rake made of wood. While these rakes can still be found at flea markets and farm sales (and new), they command a pretty high price for something so old, plus a hay rake is an easy project that can get you started making tools. Most of the old hay rakes have heads that are 24–36 inches wide with 4 to 5-inch-long wooden teeth spaced 3–4 inches along the head. The handles are usually fastened to the head with a combination of mortise-and-tenon joints and some form of bracing. We built one using a simple mortise-and-tenon joint and wire bracing to keep the head and handle perpendicular, but there are endless possibilities when making rakes.

Some old hay rakes have thin pieces of wood bent into arcs and hand-riveted

Fig. 6.1: *The extra long tines on the hay rake displayed here are particularly useful for raking lighter crops and straw.*
CREDIT: OSCAR H. WILL III

to the handle and head to make the brace; others show the end of the handle split and bent into a "Y" with the legs of the "Y" mortised into the head, about 8 inches apart. Whenever the handle is split that way, a forged iron ring, a wrap of steel or copper wire, or other clamp is installed above the end of the split to keep it from migrating further up the handle. One of the most primitive hay rakes we've seen utilizes an almost symmetrically forked sapling, with the tips of the fork mortised into the rake's head and pinned. Use what you have, use your head, and have fun casting detailed instructions to the wind as you create lovely, functional tools of your own.

All you need is access to some saplings, a drawknife, froe, or other means to split and whittle rake teeth, and some boring and cutting tools to shape and fit the handle to the head.

Old-time Hay Rake You Can Make

Some time back, we thought it would be fun and good exercise to make some hay the old-fashioned way. We had scythes and a really old antique hay rake. But one day, the modern world collided with the old — literally — the driver of a utility vehicle accidentally ran over the old rake, smashing it to bits. Loss of the hay rake led to a weekend excursion to a number of regional antique and "junk" stores in search of a replacement. Although the day ended up being what we call "a grand day out," who knew that in our part of Kansas, old hay rakes were so expensive! And then it hit us. If our ancestors could make do with what they could make, then so could we.

The initial quest was for a set of proper hay-rake plans. After hours spent scouring old woodworking books and various websites devoted to old agricultural ways, I (thankfully, as it turns out) found neither complete step-by-step plans nor any particular design so compelling that we wanted to copy precisely. Instead, with rough dimensions in mind, we headed out to a small copse of woods that grows near the center of our farm. That location is home to many tree species, including some good-sized American elm specimens (that I left intact) and plenty of hackberry saplings.

From that copse, I harvested a single hackberry sapling that was about 2 inches in diameter at its butt end, tapering to about an inch at the canopy. This little tree was roughly 10 feet tall. Once the branches were chopped from the trunk we had a piece about 7 feet long that would become the future rake's handle. I used a machete for all of this work, but you could easily use a small hatchet or bow saw with good results. A chainsaw would be overkill, but if that's what you have, go for it.

Fig. 6.2: *Three easy-to-construct hay rakes. The author describes making the rake on the far right.*
CREDIT: KAREN K. WILL

Next, we headed to a pile of dead trees to harvest the remaining wood for the project. The trees had been dozed into a pile the previous winter and were still quite green. Using a chainsaw, I cut a roughly 30-inch-long piece of 6-inch-diameter hackberry for the rake's head and a 20-inch-long, 8-inch-diameter piece of American black walnut for the rake's teeth. I decided on walnut for the teeth because it's denser than the hackberry, and I thought it would look great.

With all of the pieces back at the woodworking shop, I went to work fashioning the rake. The first step was to strip the bark from the hackberry sapling and smooth it for the handle. I accomplished this by carefully clamping the sapling in a bench-mounted vice and working with a drawknife and spokeshave. (This task is more easily accomplished using a shaving horse to immobilize the material, but while we have a shaving horse on our list of homestead tools to make, I've yet to get to it. However, lack of the best-tool-for-the-job has never really gotten in the way of our progress.)

Next, I shaved one of the handle's ends into a mildly-tapered rectangular tenon roughly 4 inches long with a cross section 1½ inches by 1 inch at the end. Since the handle was fashioned from a green sapling, I set it aside for about a week to dry. It didn't shrink much, and it never cracked significantly, although it wouldn't have been surprising if it had.

The next step was to fabricate the rake's head. This is an area where you want strength and light weight. I resisted the urge to overbuild it — hackberry is plenty strong and dries light. Using an antique splitting tool called a froe, I rived (split) out a blank that was about 30 inches long, 1½ inches thick, and 2½ inches wide. Shaping and smoothing was accomplished with a drawknife, handsaw, and a flat-soled spokeshave. I left it about 2½ inches wide in the center, where the handle attaches, but tapered it to a bit more than an inch toward the ends. In the photo, you can see that the shoulders were sawn and the ends trimmed with the saw. The final piece is just over 24 inches long. You can make your hay rake as wide as you want, but don't go too wide; thirty inches is plenty if you plan to rake heavy swaths.

The next task was to make the rake's teeth. Using the froe, I rived out roughly ⅝-inch-square lengths of black walnut from the 20-inch-long log, which were rounded using the drawknife and spokeshave — working the wood down to "dowels" a shade more than ½-inch in diameter. I cut the dowels into thirds for six of the teeth and then cut three teeth that were about 9 inches long to use where the head was thicker. A bit of whittling on one end of each tooth with a sharp pocketknife rendered them bluntly pointed (this is an important step, as it will help keep the teeth from splitting as they are dragged across the ground).

Next, I traced the handle's tenon onto the thick center section of the rake's head and, using a handheld drill and chisels, cut a through-mortise into the rake's head, taking care to taper it slightly, so that with a few whacks of the mallet, it would fit snugly to the handle. Once the mortise was cut, I whittled on the tenon a bit to get the right fit. With the handle and head fit together, I carefully bored a ½-inch hole through the head and the handle's tenon and drove one of the long teeth through the hole and out the bottom until it extended about 5 inches. That double-duty tooth pegs the handle to the head and helps gather the hay.

After marking the locations for the remaining six teeth, I bored the remaining ½-inch holes and installed teeth by driving them into place with a mallet. Through nothing short of a miracle, the teeth were sized just right — the head neither split as it dried, nor did the teeth fall out after a couple of years of use. The final steps were to trim the tops of the teeth flush with the rake's head (except for the center tooth), give the handle a light sanding, and install an angle brace in the form of twisted wire to keep the rake's head perpendicular to the handle.

Raking handmade hay is even more rewarding when you do it with a homemade tool. Making it required a little ingenuity and a bit of labor, but it was worth it. It feels just right in my hands.

Wood Forks for Pitching Hay

Once you've got your homemade hay rake in hand, you can round out your hay-gathering hand tools by creating your own pitchfork. We tend to use old metal three-prong forks (purchased at antique stores for a few dollars each) when making hay by hand, but wooden forks are pretty straightforward to make. They are light, simple to use, and can be made for free!

The most primitive, and quite possibly the easiest, pitchfork you can make requires little more than spending time among your wooded acres in search of saplings or branches of sufficient scantling to offer a handle that's already attached to three or maybe four small lateral branches coming from the same node. Think of a "Y" with a third arm located between the principal two. Look for fairly strong hardwood for this project, although cured cottonwood forks are both lightweight and strong enough — if you don't try to set any hay-pitching records.

Once you've found your sapling, cut it and trim the trident end longer than

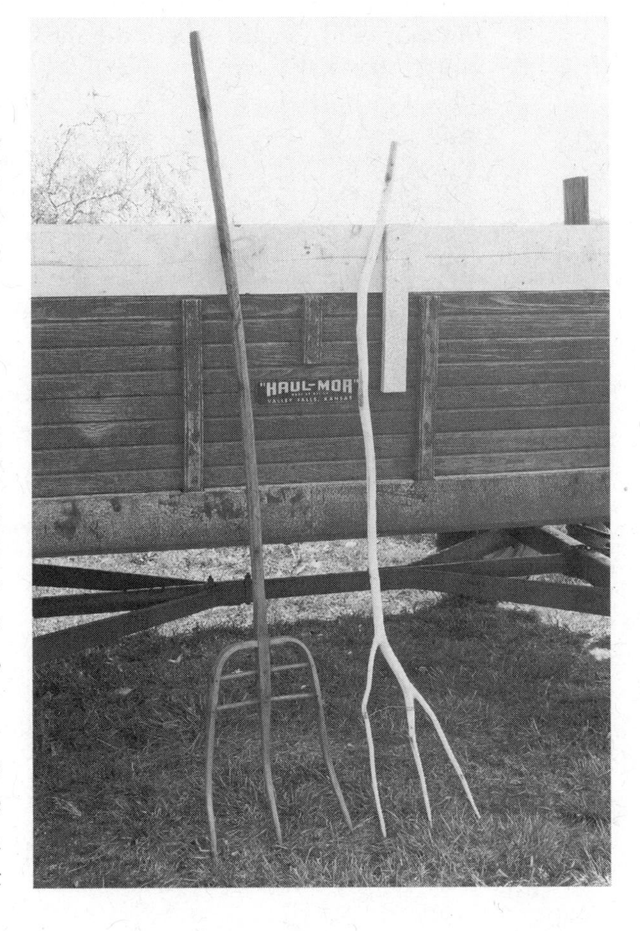

Fig. 6.3: *A sophisticated and very stout pitchfork (left) and a very primitive style. Both are entirely functional and well within the realm of the homestead woodworker.* CREDIT: KAREN K. WILL

you need it to be. Strip the bark, shape the handle, rough-shape the tines, and set it aside in such a way as to cause the three tines, to curve or bend in one direction. Some makers will actually clamp the tine end in a form of sorts to encourage the tines to dry in a pleasing and functional curve. Once the tines appear cured — they should spring back to their original shape some, but not entirely — go ahead and remove them from the form and give them a final shaping. Point the tine ends and smooth their lengths to get the most out of this tool.

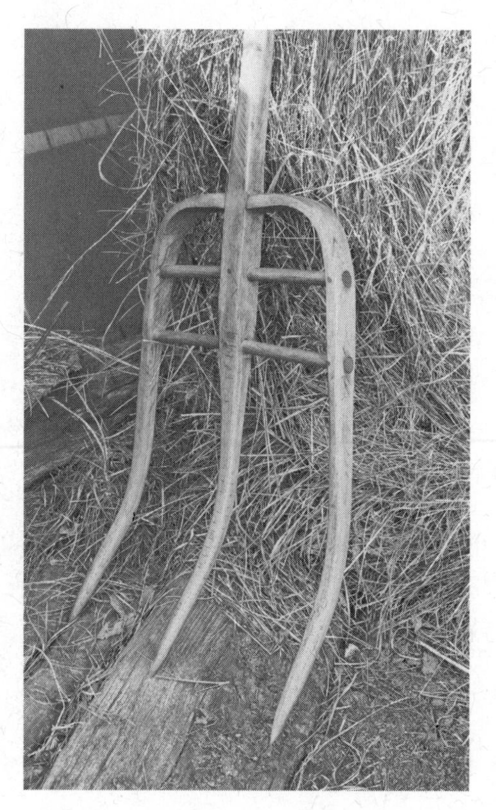

Fig. 6.4: *This fork was made by steaming the tines and bending them into shape. The edge tines were then tenoned into the handle, and the works was doweled to keep the tines oriented properly.*
Credit: Karen K. Will

If you've got plenty of fairly straight-grained saplings such as ash (though walnut and other species can work too) around your place, and you aren't afraid to try your hand at bending green wood, you might try making a fork with tines that you split and shape from one end of the handle. Your first step is to harvest a straight, relatively knot-free sapling that's about 3 inches in diameter at its butt end.

Once you've stripped the bark off, shape the handle with a rounded cross section down toward the butt end and then flatten it to 1 inch thick, about 2 feet from the end. From this flattened transition toward the handle end, size the handle to fit comfortably in your hand. Allow the material to swell to its natural width where the flat section begins and extending to the butt end. Next, set a rivet through the handle or install a band clamp around the handle about 2 inches (toward the butt end) from where you created the transition from flat round to flat.

With the piece turned on its side, bore a ¼-inch-diameter hole from one edge of the flat section through the other, about 6 inches from your rivet or clamp. Next, carefully make two splits from the butt end up to the rivet or clamp; use a chisel, small wedges, small froe, or whatever you have on hand to create three tines. Once you have the splits created, you can spread them by inserting a dowel through the hole you bored earlier and pinning the tines in place.

The next step involves tapering, shaping and smoothing the tines. Finally, you need to create a nice curve by clamping the tines in a jig. Ideally, you'll want to actually steam or boil the tines for about 30 minutes before clamping them in the jig to get a good bend. Forks with up to five tines can be created this way (you'd make three splits).

There are many other ways to make wooden pitchforks than the two described above. One of the forks in our collection uses heavier scantling material with outer tines that have shoulders bent 90-degrees and mortised into the handle; the handle itself terminates with a contiguous middle tine. This fork's tines have all been bent into a curve toward their tips as well. We've also seen pitchforks created along the lines of the hay rake discussed above. In this case, the head was narrower and somewhat stouter, and the straight tines were set at about a 10-degree angle to the handle.

Fig. 6.5: *We fabricated this garden harvest basket using scrap pine and walnut left over from the kitchen island project described in Chapter 5.* Credit: Karen K. Will

Bring in the Harvest

There comes a time in the life of virtually every homestead when a good old-fashioned harvest basket is precisely what's needed — and we don't mean the plastic junk you can find virtually everywhere. We're talking about the wooden-slat or wire-mesh baskets that make harvesting vegetables a breeze. And the best part about making these for yourself is that you can make them straight from the woodlot, with scrap lumber, or some combination of woodlot, lumber, and hardware. Let's build a basket with slats, first.

The wooden-slat harvest basket or box can be constructed in infinite ways. Here we give you one method with some alternatives, but let your own creativity, experience, and available materials guide your process.

First, decide on the basket's end material — dimensional lumber or rived planks work nicely for these pieces. The plan given here can be made from scrap material that's 9 inches by 12 inches long, and about ½- inch to 1-inch thick — but feel free to modify.

The next step is to mark and trim the bottom corners of the end pieces so that they have a 45-degree angled edge about 2 inches long (you can skip this, if you don't want the illusion of a rounded bottom — likewise, you can make ends with the sides and bottom forming a gentle curve.) Quarter-inch material works nicely for this basket's sides and bottom. If you have a bandsaw, you can mill this lumber from thicker stock. Rip two slats to 1¾ inches wide and sufficient additional slats to fill the 5-inch vertical sides and the bottom, and trim them to 18 inches in length. You could also simply harvest green wood and rive out sufficient material for the slats — these slats would be stronger than sawn slats.

You don't have to use any glue to attach the slats to the end pieces, but you most certainly could. If you don't use glue, you'll want to use fine ring-shank nails — two or three per end — because they resist pulling out. You can find these nails in the siding section of the fastener department of your local home improvement center. Drill very thin pilot holes for these nails to minimize the chance of splitting the slats; drill the holes at angles so that the nails' points will face one another when driven into the ends. This further helps keep the slats soundly attached.

Once the basket is evenly sheathed with well-spaced slats, cut two uprights roughly ¾-inch by 1¾-inch in cross section from material similar to the ends (or rived from greenwood), bore one of each of their ends to receive a handle, and nail them to the basket's ends. If you choose longer nails than the uprights, and the basket ends are collectively thick, you can turn the nail back into the inside surface of the ends, thereby clinching the uprights in place.

Now head out to the woodlot and source a branch or sapling suitable for the basket's handle. Strip the bark and shave the handle ends to fit snugly in the holes you bored in the uprights, and pin them in place. If you don't want to drive a pin through the uprights, simply shape the handle so that sufficient material extends to the outside of the uprights, bore a hole through the handle, and whittle a dowel to fit. Depending on how your handle is sized relative to the holes in the uprights, you may find that it makes more sense to install one upright, insert the handle into it, then insert the handle into the other upright, and, finally, attach that upright to its end of the basket.

The most common variation on this garden basket foregoes the slats altogether and replaces them with galvanized wire mesh (you also could substitute light-gauge galvanized sheet metal). In this case, the ends are typically formed with curved bottom corners — use your coping saw or a powered jigsaw or band saw to get the shape. Next cut a pair of stringers as long as you want your basket to be. These stringers should be about as thick as your end material and at least 1½ inches wide. You could substitute a pair of relatively straight branches or sapling sections for these pieces.

Attach the stringers to the basket ends at the top outer corner — you can just nail them to the top, or notch the top so that the stringers will be flush with the outside surface. Install uprights and handle as before, and then cut and position galvanized hardware cloth (¼-inch or ½-inch mesh works well) tight to the bottom's curve, and staple it to the end pieces. Don't spare the staples if you intend to put a lot of weight in the basket (which you probably will!).

Dibbles and Drills

There's little doubt that one of the earliest planting tools was essentially the same as what we call a *dibble* or *dibber* today. A brief study of agricultural practices among native North Americans reveals that antler tines and fire-hardened pointed sticks were used for working ground prior to planting, as well as for making holes in the earth for planting seeds and transplanting seedlings. By dragging those pointed sticks in the earth to create a shallow groove or drill, you'll see how the concept of row crops got started.

A quick search of the Internet shows that dibbles are still in production — most are wooden, complete with steel-clad tips — some have nice handles on their ends. As lovely as they are, they're beyond simplistic, so it's difficult to imagine spending $5 on a dibble, much less more than $50 for a top-of-the-line model. If you enjoy whittling, you can make dibbles all day long — and in different lengths and diameters. You can even carve fun and comfortable handles into their ends or experiment with mortise-and-tenon joinery to give yours a nice cross handle at the top.

Get started by heading out to the woodlot to source some material; a 1½-inch-diameter branch would be a great starting point. Strip the bark, cut the branch into 8-inch-long pieces, whittle and shape until you have a stick that feels good in your hands and has one end rounded and the other tapering to a point. If your ground is sufficiently hard, you can extend the life of your dibble by slowly charring the tapered point — don't let it catch on fire, but cook it until it has a thin layer of char.

If you do a lot of planting with dibbles, you might want to cut notches or grooves into it to mark specific depths.

The Comfort of Projects

I don't lose sleep over being broke one day; however, I do lay awake at night brooding about running out of raw materials for my various projects. Materials like lumber, scraps of wood, metal roofing, wire mesh, antique machinery, and the like are like money in the bank for me. As long as I have raw materials to convert into useful, unique objects (at the exact time inspiration strikes), I have no worries. That's why it's important to nab these seemingly random materials when you find or uncover them — not when you actually need them. Otherwise, that project may never happen. Oh, and try to keep it all organized, otherwise you'll end up on an episode of "Hoarders."

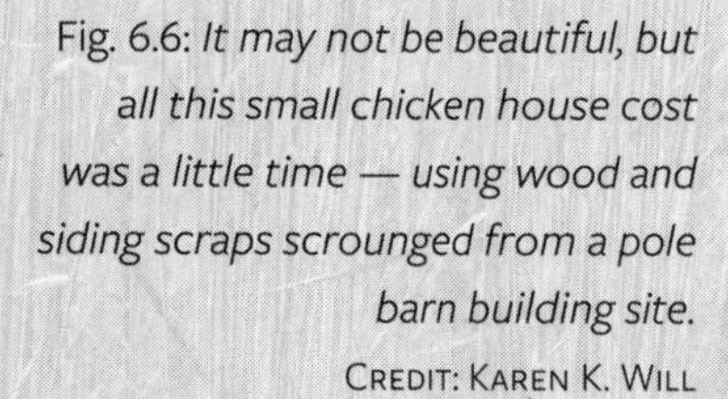

Fig. 6.6: *It may not be beautiful, but all this small chicken house cost was a little time — using wood and siding scraps scrounged from a pole barn building site.*
CREDIT: KAREN K. WILL

Fig. 6.7: *Range shelters for your broilers and laying flock will invariably need to be homemade. This functional example uses a combination of new and recycled material. If you don't like the weathered look, feel free to slap a coat of paint on it.*
CREDIT: KAREN K. WILL

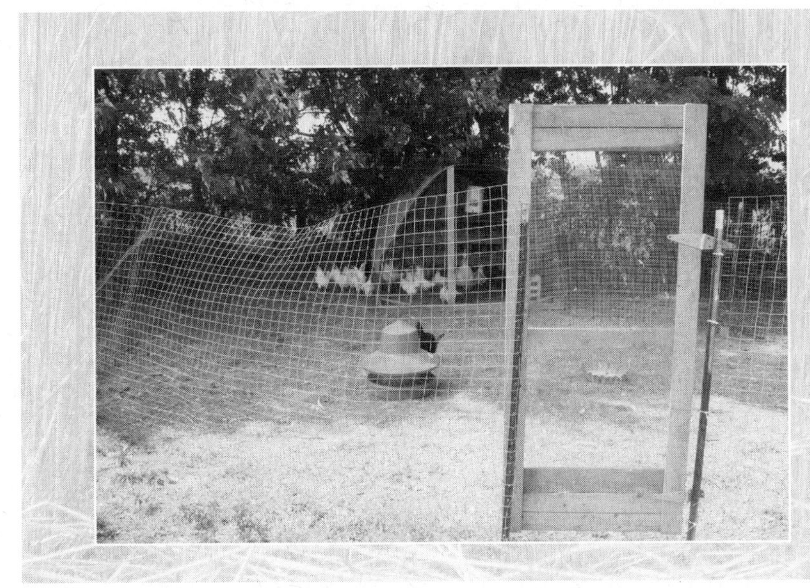

Fig. 6.8: *Temporary broiler quarters that makes use of an old screen door, some welded-wire fencing and a tarped shelter in the background. It may not be pretty, but it works — and you'll only need to look at it for about 12 weeks.*
CREDIT: KAREN K. WILL

And, if you do a lot of transplanting or sow relatively small seeds, you might experiment with dibbles of different diameters. In any case, you'll certainly be well served by homemade dibbles, and you can use the same general model to create longer stakes, digging sticks, campfire pokers, and even primitive spears.

While there's absolutely nothing wrong with using an elongated dibble or digging stick to create drills for planting, if you're trying to keep your rows fairly even, or even if you just want to mark rows in the soil (as opposed to using stakes and string) for transplanting onions or the like, you can create a garden drill quite easily. If you make it a two-row drill, you can keep all of your rows parallel.

This two-row drill is an adaptation of a 17th-century animal-pulled sled with runners set a row's distance apart from one another. This implement was pulled through tilled ground with varying amounts of weight on it depending on how deep the seed or sets needed to be planted. This drill is suited for smaller spaces and finer seedbeds.

First, construct the runners of your drill from material that's no more than 1 inch thick. Start with rectangles roughly 5 inches wide by 2 feet long, and shape them with a slope at the front, like a sled runner. Using your hand planes, drawknife, and spokeshave, shape the runners' bottoms into a curved "V" — run this shape up the sloped section at the front.

Next cut three cross members (or two, or one, depending on material on hand) to a length that's one runner thickness longer than the distance you want your rows to be from one another. Adding the runner thickness will place the runner centers at the correct distance. Attach the cross members to the runner tops using screws and glue — be sure to position the runners parallel to one another, and be sure that the cross members and runners are perfectly perpendicular.

Using the drill is as easy as attaching a loop of light line to the runners and pulling it through your tilled garden. You can weight it for deeper grooves using bricks. This method of creating drills in the earth can help you conserve seed, which will be important if you plan to grow any small grains down the road. It's also invaluable for getting difficult-to-cultivate crops spaced properly. If weighted sufficiently, it can even be used to open little furrows for planting carrot sets or plants.

How-to Hurdles

Nowadays, we call them livestock panels (sometimes, *cattle panels* or *hog panels*), and most are welded up from tubular steel. They are the modern version of lovely wooden panels called *hurdles* used in medieval England. Make yourself a set of these to enclose your garden or to keep your sheep contained. Whatever use you put them to, making a set will definitely give you the hang of working with green wood, and you'll learn the nuances of riving in the process. You could also substitute sawed lumber for most of the pieces, though the end result won't be quite as strong.

The first step is to settle on your hurdles' dimensions. For the sake of example, let's say that you are making 6-foot-long hurdles that are 42 inches tall, with five rails. We'll also add a

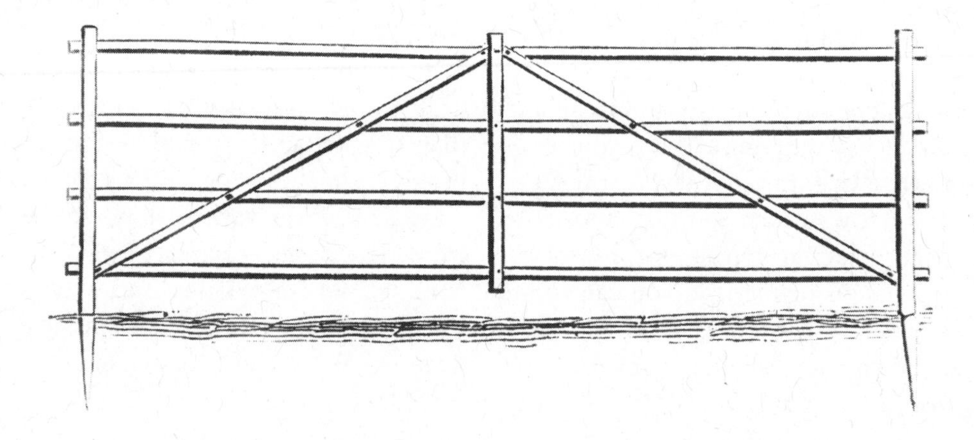

Fig. 6.9: *Hurdle schematic patterned after a late 19th century drawing. The rails are mortised into the vertical standards at each end with the center and diagonal braces riveted or clinch nailed to the rails. Add rails to suit your needs. (Note the sharpened standards.)* CREDIT: OSCAR H. WILL III

vertical brace in the center and a pair of diagonal braces on both sides of the vertical. The first step is to obtain a pair of hardwood standards (the vertical structural members that support the rails; they could be made of straight-grained oak, Osage orange, or even hackberry or walnut) that are about 1 inch by 2½ inches in cross section and 42 inches long. You can rive these from straight logs that are at least 8 inches in diameter. (If you go this route, don't worry about thick or thin spots, and feel free to dress the surfaces with a drawknife.) At the same time, you can rive out a ¾-inch by 1½-inch vertical brace; but don't cut it to length until you have the rails in place. You can also rive the diagonal braces, but take them from a log that's several inches longer than the standards — 48 inches will probably be sufficient, but add a few more inches if you can. Finally, rive out the five rails (½ inch to ¾ inch thick and about 2 inches wide should work fine). Trim these to 72 inches long.

If you decide to use sawed lumber instead, mill the logs to the same thicknesses as above and cut the standards from good, clear stock.

Traditional hurdles have standards that are pointed at one end. You can reproduce this feature with a cooper's axe or single-hand broadaxe quite easily. Alternatively you can form the points using a saw, spokeshave, or even a large sheath knife. Whichever you choose, shape the rail's end to a rough dimension of ⅝ inch thick by 1⅜ inches wide, and round off the corners. The shaped part will be tenons that will fit into through mortises cut into the standards.

Lay out the arrangement of the rails as you want them on the standards and mark the locations for the mortises. You'll want to cut ¾ inch by 1½ inch mortises — use a ¾-inch ship auger and a bit brace to define the ends of the mortise and to bore out the center. Remove the material between the holes with a chisel.

Insert the rail ends into the standards and pin them in place as follows: first the bottom left and then the top left. Next, measure the diagonal dimensions of the hurdle and tweak until they are the same — this ensures that the hurdle has a chance of being square. Next, pin the top and bottom rails into the right-hand standard — check the square again. Measure the vertical distance between the top and bottom rails, and cut the center brace to length. Locate the brace at the centerline of the hurdle, and nail it to each rail — use longer nails than necessary and clinch them as with the harvest basket handle uprights described above.

Check the square once again and tweak, if needed. Now you can trim the diagonal braces to fit from the top rail adjacent the middle brace to the bottom rail adjacent the standard. Nail these braces as with the center brace. At this point, your hurdle should be pretty rigid — go ahead and pin the remaining rails in both standards. Congratulations, you've just created something very useful that can be modified into a gate, trellis, hay-feeder side, hay-drying rack (take two and wire them together as an A-frame and drape hay over it), or you name it.

If you make a set of ten or so hurdles, you can make wire loops or bails to attach them to one another, yielding a highly flexible enclosure system that can be configured in many different shapes. And if you go to all the work to make several of these useful devices, you might want to go ahead and build a rack for them, so you can hang them off the ground when not in use. Mount the rack on wheels, and you have an incredibly easy-to-use portable corral system that didn't cost you much more than a few nails and a set of wheels.

Mudroom Bench

Some years ago, we tore the old mudroom off the house and rebuilt it ourselves, expanding the footprint to include a laundry/utility room, mudroom, and pantry. Shortly after we completed that project, we found ourselves in need of a bench where we could put on and take off our boots. Since handmade is always preferred over store-bought, we set out to build a rustic bench using hand-worked walnut.

In this case, we already had a few huge, lovely slabs of black walnut that we purchased dirt cheap from a sawyer in Missouri. (The time to buy these things is when you see them, even if you don't know what you'll do with them at the time.) Our first task was to hoist the 200-pound slabs onto a pair of saw horses to assess their grain and the character of their live edges. Then we drew out the rough dimensions of our planned bench. The slab we chose was sawed from a large crotch using what must've been a giant bandsaw mill — it was pretty uniformly 2½ inches thick and contained both sapwood and heartwood figuring. One edge was more or less straight, so I cut the 30-inch-wide piece down to about 16 inches wide on the table saw (a circular saw would also have worked) and trimmed it to about 4 feet long.

Next, I used hand planes and a power belt sander to knock down the bandsaw marks created by the sawmill, to smooth the edges, and round some of the sawed corners that weren't naturally rounded as part of the tree. Then I turned my attention to the legs. After a bit of consideration, I decided to forego the ripping of dimensional legs from another of the huge slabs on the table saw and instead selected a roughly 20-inch-long by 14-inch-diameter walnut billet that was lurking in

Fig. 6.10: *With a little ingenuity and a few hand tools, you can convert a found slab of beautiful hardwood and a few pieces of firewood to something as useful as this mudroom bench.* Credit: Karen K. Will

the corner of the workshop. I had cut that log from one of the snags on the farm about nine months earlier, so it was drier than is ideal for splitting and riving, but I used it anyway.

I had to work a bit to make the first split with the froe and maul, but the subsequent splits were like butter. In short order, I had four beautiful leg blanks that tapered from about 3 inches square at one end to 2 inches square at the other. After about an hour of shaping with the drawknife, spokeshave, and a sheath knife, I had four nice rustic legs with a mostly round cross section. I sanded them lightly, out of habit more than need.

Traditional methods would have had me cut tenons on the upper ends of the legs and let them into angled mortises in the bench top. Just as I was about to embark on that exercise, four 6-inch-long lag bolts surfaced on my workbench, so I decided to use mechanical fasteners this time instead.

I proceeded by trimming the top and bottom of each leg to parallel 11-degree angles using the table saw. This work could easily be done with hand saws, and, in hindsight, it might have made more sense to trim the ends while the legs were still more-or-less square in cross section. Happily, my approach worked, but it probably wasn't the best way. Next, I located mounting holes on the bottom of the walnut slab, and used a ship auger to bore holes that angled about 11 degrees toward the corners.

Once the holes were bored, I chopped them roughly square where they pierced the bench's top and deep enough to enclose the lag-bolt heads plus about a ¼ inch, so that I could fill the void with wood putty — the result would roughly approximate the look of a through tenon. (If you're wondering why I cut so many corners on this project, it's because I had a fairly brief time window in which to complete the project.)

Final assembly involved boring holes into the ends of the legs to receive the lag bolts, positioning the legs, and wrenching the bolts home. Somewhat incredulous at the ease with which the assembly came together, we tested the bench in the shop with our collective weight. It didn't creak, groan, or flex. Good enough!

To finish the bench off, I sanded, scraped, and whittled until everything was nice and smooth. Filled the fake tenon ends with wood putty, sanded some more, and soaked the works with warmed, boiled linseed oil. Once the oil set, I worked the surfaces with steel wool, applied another coat, and then simply rubbed it all with a lint-free cloth until it looked the way we desired.

The bench has been put to good use in the mudroom for several years now, and it's held up nicely. Even though I expected some checking (cracks at the end) and possibly even splitting (cracks within the board), so far, none have appeared. My plan is that the next time I build a bench, coffee table, or what have you, I'll go ahead and join the legs to the top using more traditional methods — hopefully, I'll have my shaving horse and assembly bench completed by then too.

Section 3

Gym Membership That Pays

As the North American physique has grown ever weaker and larger, there's been an explosion in club-style exercise programs — e.g., Jazzercise, Curves, 24-Hour Fitness, etc. Then, there's the explosion in fad diets and other magic bullets aimed right at our wallets — while claiming to give us back our collective youth and figures. There is nothing more ironic than paying money for something as simple as exercising — or exercising for the sake of exercising — when there are lots of ways to work your body for free, or nearly so. And, instead of staring at a wall or TV screen while "working out," you could be observing a lovely landscape, or even doing something completely productive and fulfilling. Further, so brainwashed have we become that we'll drive the few miles to the gym (that we're paying to use), to use machines that we're sure will help us lose weight and stay in shape efficiently. If folks simply walked to the gym, they wouldn't even have to go in — or buy a membership! If the real reason for the modern gym's success is born of a need for social interaction, then so be it, but exercising in a gym certainly doesn't make sense in most other ways.

Perusing old black-and-white photographs, linotypes, and other illustrative relics left behind by our rural ancestors, you might notice that those folks look emaciated by today's standards. The work involved in simply living day to day was rigorous and grueling. If you read *Sod House Days,* a transcription of a late 19th-century homesteader's diary, you might be surprised at just how tough the work was. Human labor was the second most important resource those folks had (brains being the first). In *Sod House Days,* Kansas homesteader Howard Ruede thought nothing of walking 13 miles to town to check his mail and do business once a week. Sometimes he was able to secure a wagon ride at least partway home, which was a huge help when his provisions were particularly bulky or heavy. He thought nothing of walking five miles to a neighbor's house to ask to borrow a tool (oftentimes, the tool was in use, so the answer was "no") or attend a worship service. He often made the roundtrip twice in a day — if a tool's owner wanted it returned straight away, that's what he did. Today, folks think nothing of hopping in the car to go three blocks to visit a friend or pick up a gallon of milk.

Certain aspects of pioneer life were pretty tough, and drudgery does little good if it goes on forever; however, there is plenty of pride, production, and even good health buried in much of what savvy marketers would have us believe is pure drudge. We're not at all averse to employing technology and tools in our daily toils, but if all we're doing is saving time and energy so that we can drive to town to go to the gym — well, you get the point. There are plenty of ways to get your exercise around your place. Working in the woodlot is one; weeding the garden or turning the soil are others. Harvesting crops by hand, digging postholes by hand, you name it, can all be considered a gym that pays. We've experienced the mind-numbing boredom that comes from riding a stationary bicycle or running to nowhere on a treadmill — remembering those days keeps us going after an hour of weeding the garden or planting corn or making hay.

The homestead "workout" doesn't always have to be about work. We're lucky enough to steward sufficient land that we maintain a few miles of walking/running/biking trails on the place. Sure, we have to dodge cow pies and a flock of sheep now and then, but that only adds to the enjoyment. The dogs love the walk, we get to look over the animals, fences and meadows, and not only do we get needed exercise, we also benefit from the stress-lowering aspect of being up close and personal with Mother Nature. The gym can never compete with the natural world.

Chapter 7

Row Crops You Can Grow by Hand

Planting crops in highly organized rows is one method for getting the job done; planting in rows makes it easy to cultivate, mulch, and even irrigate. It works particularly well for growing large quantities. Planting and tending row crops can readily be done using a few homemade tools and a hoe. If budget allows, the work will proceed more efficiently with a walk-behind planter and wheel hoe-type cultivator. Even if you wind up using a walk-behind rotary tiller as part of the equation, you'll expend plenty of energy bringing in the harvest.

Preparing the Earth

Your row crop patches will benefit from some level of soil working in the fall or the early spring — possibly both. In Chapter 1, we discussed methods for putting poultry to good use preparing the garden for planting. In Chapter 2, we considered using hogs to glean, clean up, and till your vegetable patches.

Fig. 7.1: *Preparing ground for the garden is hard work.* Credit: Nathan Winters

Fig. 7.2: *If your soil is relatively mellow you can prepare it for planting with a broadfork — an experienced user can work hundreds of feet of planting bed in a day. Rake it out, and you are ready to plant.* Credit: Meadow Creature LLc

Either, or both, of these animal systems can do wonders helping prepare your quarter-acre corn patch for planting, and they'll make plowing down green manure or cover crops a breeze. If you have a heavy-duty walk-behind rotary tiller, go for it — you'll get a good workout horsing that iron horse around. However, if you have neither hogs, nor a large poultry flock, nor a rotary tiller, you can use a scythe (see Chapter 8) to mow down a cover crop and a turning spade to work the ground. Alternatively, you can use a broad fork to loosen and aerate soil and a hand cultivator (a three-prong device attached to a handle) to smooth the seedbed. In all cases, you'll expend plenty of calories to get the work done. Take it in stages, and work during the cool of the early morning, if possible. Try out different methods. We find the combination of hogs and a wheel hoe (with tine attachment or a rotary tiller) tough to beat for preparing large patches for planting.

Some schools of gardening eschew regular tilling of any kind. In these systems, you might loosen soil with a broad fork and then simply plant patchwork style after pulling back sections of more-or-less permanent, deep mulch. These systems can be highly productive, although we've yet to figure out a good way to adapt them for large corn (or other grain) patches. It might be that we have the value of rows ingrained into us at birth, or it might be that rows are simply efficient. On the other hand, we do find that our little-worked vegetable garden ground stays very soft and fluffy and is easy to transplant into.

Once the ground is worked, and you've left it long enough to become friable and mellow, it's time to do some leveling and clod or rock removal. The metal "dirt" rake works well for this task, although some folks pull a couple of steel fence posts or other pieces of scrap across the patch to smooth things out. If you are going to plant your crop in ridges so you can use gravity to help with flood irrigation, now is the time to mark the rows, then hoe or use a middle-buster type attachment on the wheel hoe to cut troughs while forming ridges. If your soil is quite mellow, light, and soft, you can get much the same effect using a piece of 2 x 10 or other lumber of a similar dimension. Make it pointed at the front like the prow of a ship, add sides to it, put some weight inside, and you're ready to pull it through the garden. Likewise, there are attachments for some rear-tine rotary tillers called hiller-furrowers that will also do the trick. If you plan to simply plant the ground without furrowing, so much the easier.

Sowing Your Seed

Early sowing tools included the same sharpened sticks, antler tines, and even flaked rocks that were used to work the ground, much like we use a modern dibble. For some crops, those sticks or antler tines were used to open up shallow furrows (drills) into which seed was dropped more hastily and heavily to yield thick stands. Both methods work well for planting corn.

If you want corn in nice straight rows, try pulling a homemade drill (Chapter 6) and, walking along the furrows it forms, drop seed at regular intervals. Once the seed is placed, you can use a hoe, your feet, or a drag made from a 2 x 4 down the rows to cover the seed. Walk along the rows or roll the field in some way to press the soil into contact with the seed. If you want corn planted in hills set on a grid or in rows, mark the rows with the drill

Fig. 7.3: *Walk-behind garden seeders come in many different styles. Check for antiques at farm sales and rural estate auctions — or in your neighbor's barn!* CREDIT: OSCAR H. WILL III

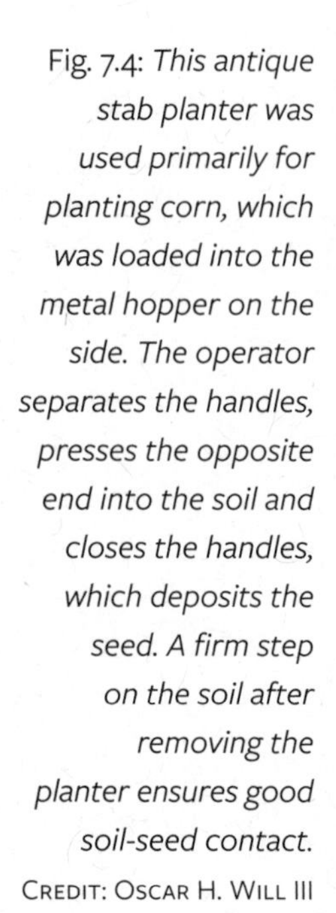

Fig. 7.4: *This antique stab planter was used primarily for planting corn, which was loaded into the metal hopper on the side. The operator separates the handles, presses the opposite end into the soil and closes the handles, which deposits the seed. A firm step on the soil after removing the planter ensures good soil-seed contact.*
CREDIT: OSCAR H. WILL III

or stakes and string, hoe up hills, and use a dibble to poke holes for planting the seed. Use your hands or the hoe to cover them and press the soil firmly into contact with the seed.

Another fun method for placing corn seed in rows or hills is using the old-fashioned stab planter. The stab planter is more complicated than the homemade drill or planting stick or dibble, but it's quite a bit simpler than walk-behind units in that it consists of an opener/seed delivery tube and some kind of seed-metering capability (sometimes it's as simple as the operator dropping the individual seeds into the tube). Closing and pressing are generally taken care of by the operator's foot. It's so simple! Much of corn country was planted with such devices shortly after the sod was broken. Generations ago, the Oscar H. Will seed company planted acres of corn in North Dakota using a team of men armed with seed-metering stab planters.

The stab-style planter will save some wear and tear on your knees and back, save seed, and reduce the need for thinning. Antique semi-automatic units can be found at farm sales, junk shops, or rustic antique stores. These devices usually consist of a hinged tube made out of wood and metal with a seed box on one side and a perforated slider that takes two to three seeds from the box and drops them down the tube and into the ground. You basically grab the two handles, pull them apart, stab the planter into worked ground, push the handles together, and pull it out of the ground. A light brush and step with your foot seals the deal. New versions of this planter include the Stand 'N Plant Standard Seeder, which can be used to plant individual seeds and small plants like onions. You meter the seed or onion plants by hand, but you just need to walk down the row to get them into the ground. The Stand 'N

Plant seeder also is capable of planting in plastic-covered beds, with infinite variation in row and seed spacing.

If you prefer a bit more mechanization and wish to plant in rows, you may want to upgrade to a walk-behind garden seeder.

In today's terms, a garden seed planter is a precision machine that places individual seeds at a specific spacing along a row. As the planter moves along the row, it opens the soil to a specific depth, places the seed, covers the seed, and provides some means for pressing the soil into contact with the seed. Walk-behind planters generally have a wheel in front to drive a seed-metering mechanism; an often-hollow, wedge-like structure called the *shoe* opens the soil and helps convey the seed to the soil, and a closing device pulls the soil back over the seed. A press wheel at the rear ensures good seed-soil contact, which is needed for efficient germination.

Working with hand planters can be joyous or frustrating, depending on your soil type, soil conditions, garden size, and your physical condition. Lighter-duty planters tend to work better either in lighter soils or in heavy soils under ideal conditions (perfect moisture content, completely mellow, friable crumb, etc.). If soil gums up on the planter's parts or is so tight that the openers can't do their job, it's best to put off planting to another day. (Instead, it might be a good time to work on conditioning the soil.)

Earthway's 1001-B Precision Garden Seeder (about $125, well-equipped)is a good starter planter. This garden seed planter has been on the market (in various iterations) for decades. The Earthway is made with lightweight aluminum and plastic components that have proven durable under most homestead circumstances. Some folks reinforce the handle structure when rivets loosen up over the years, but, overall, the planter is simple to adjust, simple to use, and can be had with seed-metering plates that work for just about anything you would direct-sow in rows in the garden — including corn seed of various sizes. If you want to change the spacing, you can simply tape over some of the holes on the plates. These days, you also can order blank plates from the company to accommodate your own custom sizes/spacing. We've planted acres over the years with the Earthway; you might choose this planter for gardens or corn patches up to about a quarter-acre in size.

For gardeners with more ground to plant, the Cole Planet Jr. push seeder (about $600, well-equipped) is a good choice. This plate-type planter is constructed of steel, cast iron, and wood (the seed box is plastic) and is based on a venerable old unit-planter design that is sufficiently stout to mount on a tractor's toolbar for multi-row, medium-scale planting. This planter isn't ideal for the smallest gardens — you really need to load its hopper with more than a packet of seed for best results, but if you have an eighth of an acre of corn to plant along with mangels, beans, and many other crops, this tool has the heft to get it all done today and be ready for more tomorrow. It's really well built; you should be able to hand it down to your gardening grandchildren. We've used the Cole Planet Jr. extensively to sow corn, fodder beets, and a number of other crops. The machine's handles are adjustable to suit different-sized people, and its weight offers great momentum once you get it rolling. As you would expect with a professional-grade tool, this planter tracks well, and its row marker doesn't skip.

The Hoss Tool Company, owned and operated by a pair of passionate gardeners, makes a small-scale garden planter as part of a truly versatile hand-gardening system that includes a wheel hoe with several different plow and cultivator attachments. The planter attachment connects to the wheel hoe using the same mounting holes as the cultivator tines and includes a rear press wheel that also drives the seed-metering plate. The unit comes with a number of pre-drilled seed plates. Blank plates are also available for custom seed-size or spacing. This beautifully crafted unit is a perfect planter option for folks who

Fig. 7.5: *An original S.L. Allen Co. Planet Jr. seeder that's still on the job. This unit also has cultivator tines and hoe knives that can be bolted to the frame once the planter unit is removed.*
Credit: Oscar H. Will III

already own a Hoss wheel hoe, or who intend to add that tool to their shed in the future.

If new isn't in your budget, look for one of the nearly thousands of models of antique walk-behind planters that still turn up at farm sales and antique stores. The key to being able to work with the antique models is to be sure that their seed-metering plates or drums or brushes are intact (or could be easily fabricated). Look for names like Cole and Atlas, though there are a host of others.

Broadcast planting of corn isn't generally recommended because it is difficult to get the seed deep enough and to keep the patch relatively weed free until the corn canopies. But by all means, broadcast your corn seed if that is the only means available — but keep in mind that you will save seed and be able to weed more easily if you choose the homemade drill method instead. We have had good luck sowing sorghum (a corn relative) using the broadcasting methods covered in the following chapters — but even then, drilling or planting with a seeder is more efficient and still provides you with plenty of exercise.

Cultivating

Most row crops will require cultivation at least twice between the time that you plant the seed and when the growing crop canopies. Once the crop canopies, the plants will tower above weed seedlings, and the shaded darkness beneath the canopy will inhibit further germination and — more importantly — robust weed growth. If you have plenty of mulch, you can cultivate once when the crop is well established — say 20 inches tall for corn — and then lay a heavy straw or hay mulch between the rows. If you go this route, it will be useful to lay the straw down in front of you, and then walk on it to pack the layer into a dense blanket. A dense layer of mulch will allow moisture to percolate down to the soil

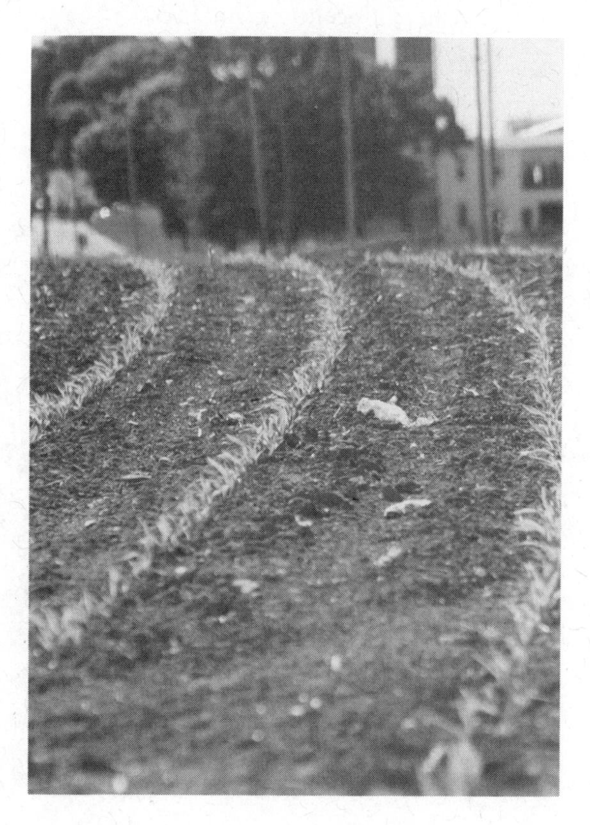

Fig. 7.6: *With a bit of luck and good timing, your direct-seeded row crops will emerge before the first wave of weeds.*
CREDIT: OSCAR H. WILL III

while causing weed seedlings to perish as they try to penetrate the mulch layer.

Cultivating row crops by hand is a daunting task — unless you have a few hand tools to help. Pulling weeds is, of course, the most rudimentary form of cultivating (except when your purpose is to loosen the soil or hill root crops; in those cases, you'll always need some form of mechanical advantage). Our most primitive ancestors once relied on the planting stick, bone or stone hoe, and antler-tine cultivators to get this work done in a reasonably efficient manner. Fast forward to the first half of the 20th-century, and it wasn't at all uncommon for folks to hoe acre-sized corn or other row crop patches using a steel hoe attached to a long wooden handle. Indeed, gangs of laborers wielding hand hoes cultivated and weeded large acreages of row crops (especially cotton and beans) in lieu of horse- or tractor-drawn equipment.

We still use hand hoes of various sorts to weed and scratch the top layers of soil — but mostly between plants within the row. To keep the space between the rows loose and clear of weeds, we use the wheel hoe equipped either with knives or with spring-tooth tines. Effective cultivating leaves weeds uprooted and the soil surface loose so that it can dry to create a mulch of sorts that will reduce weed germination near the surface. We've used a number of antique wheel hoes that we found at farm sales and junk shops. If

Fig. 7.7: *If you take care with your row spacing, cultivating the weeds is readily accomplished with a wheel hoe that has cultivator tines attached. This method is as effective as hand hoeing, and a bit quicker overall.*
Credit: Karen K. Will

you want to go that route, you definitely want to be sure that the wheel and carriage are functional and that you have at least one set of knives or serviceable tines specific to the individual brand and model. If you find a wheel hoe sans accessories — and you're good with metal fabrication — it's often possible to create the attachments you need.

Using a wheel hoe is as easy as grabbing the two handles and pushing the unit between the rows. You'll find your own rhythm, but it invariably consists of a push forward, partial pull backward, step forward, push again, and so on. Operating the wheel hoe in non-compacted ground is pure joy because it gets your heart rate moving, it's quiet, vibration- and fume-free, and it's much easier on the soil overall than a rotary tiller. You can also fit some wheel hoes with reversible moldboard plow blades, which will not only turn soil but also act as hiller/furrowers — which is really handy for some crops and/or irrigation methods.

We generally use the hogs to work corn ground, dress it up with a wheel hoe wielding tines, plant with the Cole Planet Jr. seeder, cultivate twice with the same wheel hoe setup, and then let what happens happen. Using this method, we wind up with fewer than 12 person-hours in the crop until harvest rolls around. Strange as it may seem, some of our ancestors harvested

Fig. 7.8: *For folks who want to shock their corn crop, bundles of corn are cut, bound and then stacked against one another to form a shock. Early mechanized corn binders cut and tied the bundles using horsepower — exactly as it's happening here.* Credit: Oscar H. Will III

Fig. 7.9: *Row crops such as grain sorghum can also be shocked to facilitate curing and storage through the winter.* Credit: AbundantAcres.net

acres of field corn completely by hand by snapping the dried ears from the stalk, ripping the husk from the ear, and tossing the ear into a horse- or tractor-drawn wagon. So important a skill was this style of corn picking that hand picking contests still occur all over North America today. If your corn patch is a couple of acres or less, there's no reason not to hand pick, unless you want to *shock* your corn, instead.

There are lots of different ways to shock corn. In some cases, the stalks are cut by hand with a corn knife — in others, a specially designed corn knife is strapped to the outside of one's boot. You grab the stalk, give it a kick, and move on to the next stalk. In both instances, the stalk cutter continues until there's a nice bundle of stalks — perhaps an armful — which are then bound together about a third of the way down from the top. Some folks use twine to bind the bundles; others use a flexible piece of corn stalk. Once several bundles are made, they are stacked against one another in a hollow tent-like structure, then (usually) bound all together at the same height where the individual bundles were tied.

A well-made shock is a great way to store corn for later use (for instance, when a corncrib for picked and husked corn isn't available). Many folks still remember fondly (or not so fondly) the wintertime chore of breaking shocks, removing the ears to feed the livestock, and using the leaves and stalks for bedding or roughage. For folks who pick their harvest, winter is the time to turn animals into the harvested patch to glean precious grain,

Fig. 7.10: *Shelling ear corn entails removing kernels from the cob. You can do it with your hands, but if you have a household's worth to shell, an old hand-cranked box sheller will process it in a jiffy.*
CREDIT: OSCAR H. WILL III

to consume or break down the stover, and prepare the soil for planting again in the spring.

If the corn or other grain in question was grown for human consumption, the exercise associated with using the bounty is far from over. In the case of corn, at some point you'll want to remove the kernels from the cobs, and there's no better way to do this (on a smallish scale) than with a hand-crank corn sheller. These devices range in size from the diminutive box-mounted models that you crank with one hand while feeding it with the other, to the larger, hopper-fed models that you crank with both hands. The bigger models generally have a large shelling wheel that acts as a flywheel, which allows you to crank it up and reload the hopper without the shelling coming to a halt. Believe it or not, some of these hand-powered shellers can deliver many bushels of shelled corn per hour (a bushel of shelled corn is about 56 pounds of kernels, depending on the moisture content).

Most of us aren't keen on eating flour, flint (very hard grinding corns, related to popcorn), or field corns whole, but we sure do like our cornmeal. Once you've burned the calories to shell the corn, you'll burn a ton more grinding it into a coarse polenta meal — and even more if you grind it into a corn flour. We enjoy growing all of our meal corn using tasty old flint and flour varieties such as Floriana and Mandan Bride. It takes us about 40 minutes of hard cranking to mill roughly 8 pounds using a beautifully crafted Montana-made GrainMaker mill. The folks at GrainMaker are homesteaders themselves, so they understand what we need. They produce many heirloom quality, precision tools to help you on your journey. GrainMaker mills can also be powered with electric motors. The company will even sell you what you need to adapt your grain mill to run on bicycle power.

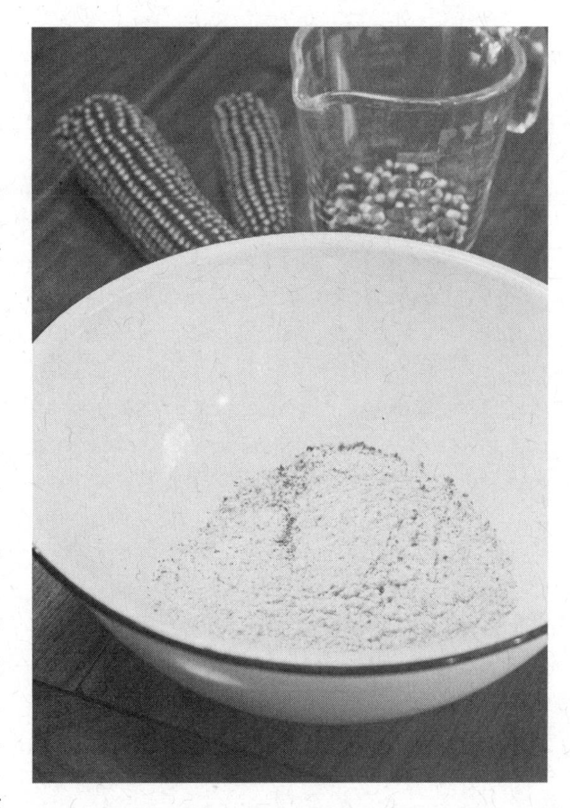

Fig. 7.11: *From kernel to meal and back again. When you grow your own, you have many different varieties from which to choose and can experience true corn flavor.*
Credit: Karen K. Will

GrainMaker Grain Mill — Hank's Pride and Joy

Each time I convert pounds of our homegrown Bloody Butcher, Floriana, or Mandan Bride corns into cornmeal with our GrainMaker hand-powered mill, I find myself covered in gooseflesh. I've used the C.S. Bell No. 2 mill to grind meal (and it is possible to make fine cornmeal with that mill using multiple passes), but the GrainMaker is truly a work of art. It delivers finished cornmeal (from coarse to fine) in a single pass. It works with other grains too; and it can whip up some terrific homemade nut butters, to boot.

Proudly made in Montana at a family-owned and operated precision machine shop, the GrainMaker reeks of American pride and a master-craftsman level of quality. With the GrainMaker, you won't find poorly finished castings or stab yourself on stray metal slivers that were missed by quality control. Instead, you'll find a perfectly functional and absolutely gorgeous piece of metal craft that is simple to use, intuitive to assemble, and works exactly as intended. Top that off with a lifetime warranty, and the mill is nothing short of phenomenal.

Although we choose to make our cornmeal using human power, the GrainMaker's flywheel doubles as a V-belt pulley, and the machine is sufficiently robust that it can be set up with motor power easily — without voiding the warranty. For the more industrious, the GrainMaker would be simple enough to drive with a bicycle or even a geared-down windmill (the company has everything you need to convert to bicycle power). The GrainMaker has a red powder-coated finish that makes cleaning it as easy as brushing and wiping the surfaces. If you mill something like peanut butter or coffee beans, though, it's a good idea to disassemble the unit and give it a good cleaning before switching over to wheat. (If you're already very familiar with mills, you may be wondering about the kind of bearings used to locate the GrainMaker's main shaft. The company installs only sealed precision roller bearings, which will give you plenty of years of hard use before they need to be replaced.)

Finally, the Jones family at GrainMaker is friendly, incredibly knowledgeable, helpful, and justifiably proud of their product. These are the businesses we homesteaders need to support — those that are true partners.

Fig. 7.12: *Hand grinding corn for flour and coarser meals is quite a workout compared with operating a motorized grinder or driving to the market — and the fresh flavor can't be beat.* Credit: Karen K. Will

Hank's Wholegrain Cornbread

Wholegrain cornbread can be the ultimate comfort food, and it makes a healthy alternative to versions made with highly processed flours and meals. I love cornbread — nothing goes better with chili, and I prefer it with minimal sugar and pure, unadulterated fats.

This recipe is based on one often printed on the cardboard tube of industrial de-germed and vitamin fortified yellow cornmeal. I substituted organic peanut oil for the "vegetable oil," homegrown and ground (fine) cornmeal for the name-brand stuff, and whole wheat flour for bleached, all-purpose white flour.

I didn't get rid of the sugar entirely, and you could substitute unadulterated whole milk for the unadulterated half & half. Feel free to experiment with substituting honey for the sugar and try other fats, such as home-rendered lard. You can bake it in a glass or glazed dish, but I am very fond of cast iron. If you use cast iron, pop the wholegrain cornbread out of the pan to cool — that way you will avoid any iron flavor in the bread, and condensation won't form between the cornbread and the pan.

Serves 6

2 eggs, beaten
1¼ cups half & half
½ cup peanut oil, divided
1 cup whole wheat flour
1 cup fine wholegrain ground corn
2 tablespoons sugar
½ teaspoon salt
2 teaspoons baking powder

Place ¼ cup peanut oil in a 10-inch cast-iron skillet and place in the oven. Preheat oven and skillet to 400°F.

Whisk together the eggs, half & half, and remaining ¼ cup peanut oil. Set aside.

Whisk together the flour, cornmeal, sugar, salt, and baking powder. Combine the two mixtures and stir just until wet.

Pour the batter into the heated skillet and bake for 20 minutes, until golden brown. Allow to cool for a few minutes before removing the cornbread from the pan.

Fig. 7.13: *With all of the calories you expend to supply cornmeal for your family, you'll definitely appreciate refueling on this delicious cornbread.* CREDIT: KAREN K. WILL

Chapter 8

Handmade Hay and Other Forages

Nothing smells like summer more than a hay field freshly mown. Even in winter, when you break into those little bundles of summer sunshine, the scent will take you back — and the nutrition will help your flocks and herds thrive. Making hay is something that is universally anticipated by folks living on the land, but for many smallholders, the expense of collecting and maintaining all the power equipment used in the modern hay meadow is just too much. Your options include buying hay, having a custom-haying crew hay your place on shares, or making what you need slowly but surely — by hand. And if you live in town and have just a few rabbits to feed, handmade hay is the only way to go.

As much as we thoroughly enjoy the sounds and smells associated with using diesel-powered equipment and modern large round balers, we've long since discovered that it's possible for us to make a portion of the hay we

Fig. 8.1: *It takes scores of thousands of dollars worth of machinery to make hay on this Big Sky Country scale. You don't have to make the investment if all you need is a few tons a year.* Credit: Oscar H. Will III

need to keep our herd and flocks going through the winter using scythes, rakes, and pitchforks. And not only do we produce feed for the animals, but we find an invigorating and healthful rhythm in the process that helps relieve stress, increases our cardiovascular capacity, and helps us sleep oh-so-well.

High Mowing Hay Ground

Ideally, your hay meadows will be free of significant numbers of woody shrubs or tree seedlings. The problem with lots of woody material is that you'll need to mow around the saplings or risk damage to your scythe. Our hay meadows contain a mixture of native and exotic grasses such as tall fescue, big bluestem, eastern gamagrass, smooth brome, and sideoats grama, to name a few. We're also blessed with a relative abundance of the tall and nonbloaty yellow and white sweet clovers, as well as a number of other native and introduced legumes — including black medic and feral alfalfa. If this mixture of tall and short, warm-season and cool-season forage sounds alien to you, chalk it up to regional differences. We live near the western edge of the tallgrass prairie ecosystem, but some of our rocky and sloped ground is most definitely representative of the shortgrass system.

If you're able to plant new hay ground, consult with your neighbors, regional hay growers, and even your local extension (or the equivalent) to get an idea of what kind of mixture you might seed. No matter the advice you receive, we recommend shying away from a monoculture meadow. The grassland matrix thrives on diversity — diversity in both grasses and forbs. It's tough to imagine a hay meadow without its share of legumes plus a combination of grasses that will produce, no matter what the season's weather brings. And it's difficult to imagine a hay meadow without a lovely array of wildflowers and aromatic broadleaf plants that make walking, and, indeed, mowing such a delight. There's something quite refreshing when your scythe slices through a patch of bee balm or sage.

Seeding hay ground can be accomplished effectively in early spring or late fall. Broadcasting the seed, either by hand or with any manner of handcranked, trailer or PTO-driven spinner will get the seed spread around. If you live in a snowy region, the best time to broadcast pasture seed is just before the first blanketing snow. The snow will moisten the soil, press the seed into contact with it, and insulate it from the bitter cold and desiccating winds.

The main drawback to broadcast seeding is that invariably the seed will fall thicker in some places and thinner in others. And unless you follow the broadcasting with a light harrowing and then roll with something akin to a cultipacker, you'll wind up with a germination rate that's less than ideal. If you don't have access to a small tractor with a 3-point hitch — and if the budget allows — you might consider renting a tractor/solid-stand seeder combination for a day or weekend. The solid stand seeder attaches to the tractor's hitch and has the means to give the soil a final shallow pulverizing to create a fine seedbed onto which it drops a nice, even distribution of seed. It then pushes the seed firmly into the seedbed with a gang of ridged packer wheels. Plant your meadow with one of these outfits just before the snow or spring rains, and you might be surprised at the lovely sward you see come July.

If you don't have access to any of the aforementioned machines, by all means broadcast the seed by hand, rake it in and roll the area or otherwise firm the soil by any means possible. And if you can't figure out a way to firm five acres of soil by hand, just spread a little more seed than recommended for broadcast application. Any seed that doesn't sprout this year may come up in the next few.

Finally, if your hay ground has any large rocks protruding more than a couple of inches out of the ground, or if there are other hard objects such as stumps, well casings, and the like, make a mental note of their location and mark them or map them so that you don't inadvertently mow into them and wreck your scythe (same advice for those using mechanical mowing strategies).

Making the Cut

If you don't already own a scythe, look for used ones at farm sales, antique or flea markets, and even online — new ones are pretty easy to find online and at specialty outlets, especially in Amish country. There are at least a couple of different styles of scythes out there, and we've used both for cutting weeds and mowing slopes. The first scythe Hank ever owned was beautiful — and so lightweight that he concluded it was a decorative replica of the real thing, so he sold it for a few bucks to a "junk" collector. It turns out that first scythe was actually a very old Austrian-style scythe. At the time, Hank incorrectly reached the conclusion that the heavy-snathed, heavy-bladed American-style scythe was the real deal — it was, after all, heavy duty. It was also really heavy — but he was dumb enough to think there was no better way, and young and strong enough not to care, so he used that scythe for many years.

The American-style scythe is a formidable tool. This scythe is generally constructed of hard, thick steel; you really must use a grindstone and file to whet it. The American-style snath is usually round in cross section and features adjustable handles that help

Fig. 8.2: *The Austrian-style scythe is lightweight and cuts well when the forage is lush and wet with dew.* CREDIT: KAREN K. WILL

Fig. 8.3: *Before cutting and whenever the scythe begins to catch during cutting, take a moment to whet its cutting edge — it'll also give you a chance to catch your breath.* CREDIT: KAREN K. WILL

you get the right fit. The combination weighs about seven pounds, assembled. You get plenty of momentum to motor through thick growth. There is nothing wrong with the American-style scythe, but once you experience the Austrian-style scythe, your old American may just collect cobwebs and corrosion — or get sold to a junk dealer. However, if your meadow has lots of saplings or other shrubby growth, the American-style scythe will be far less harmed than the Austrian style.

In a nutshell, the Austrian scythe is finesse where the American-style is brute strength. The Austrian has a pronounced crescent shape as opposed to the American's arched shape, and it's made with softer, thinner steel — relying on its specific three-dimensional shape for strength. The Austrian scythe is kept sharp by peening the cutting edge, and dressing with a curved whetstone. The peening draws out an incredibly thin, draw-hardened and sharp edge, while the whetstone keeps the edge true as a day in the field progresses. The Austrian scythe weighs little more than half a comparable American style.

Swinging any scythe effectively takes some practice — some instruction would be incredibly helpful. Take a look at the videos posted at Scythesupply.com, and read the pertinent sections of David Tresemer's *The Scythe Book*. And practice, practice, practice.

Bringing in the Sheaves

Once you have your cutter in hand, you should procure a lightweight and relatively wide hay rake — believe it or not, they are still available online and at some specialty brick-and-mortar locations. If you can't find one to purchase or just don't want to spend the money, turn to Chapter 6 for guidance on how to create one for free.

If your only experiences raking involve gathering up the wet, clumped grass that your dad's 1960s-vintage power mower belched out, raking leaves in the yard, or de-thatching the lawn, fear not. Raking hay with a lightweight wooden hay rake is heaven by comparison. It's easy and satisfying work.

In addition to the rake, a pitchfork (not a manure fork, not a garden fork, not a potato fork) designed for hay is a must. These forks can be made entirely of wood (see Chapter 6) or, more likely these days, have fine steel tines (usually 3 or 4) attached to a wooden handle. Longer handles are better than shorter handles, in general, and if your fork's tines aren't quite sharp, feel free to make them that way. A proper pitchfork will make loading hay onto your wagon and into the barn or stack an absolute joy (especially if you only do an acre or two all season).

Finally, you'll need something with which to bring in your hay. We use an old four-wheel garden wagon to which we've attached some vertical staves to facilitate the stacking of hay to about 8 feet high. We pull the wagon with a utility vehicle or garden tractor. You could use a pickup truck for this task, or do as some folks do, and just rake the hay onto a large, lightweight tarp or piece of burlap and drag the load where it's needed. You could also skip

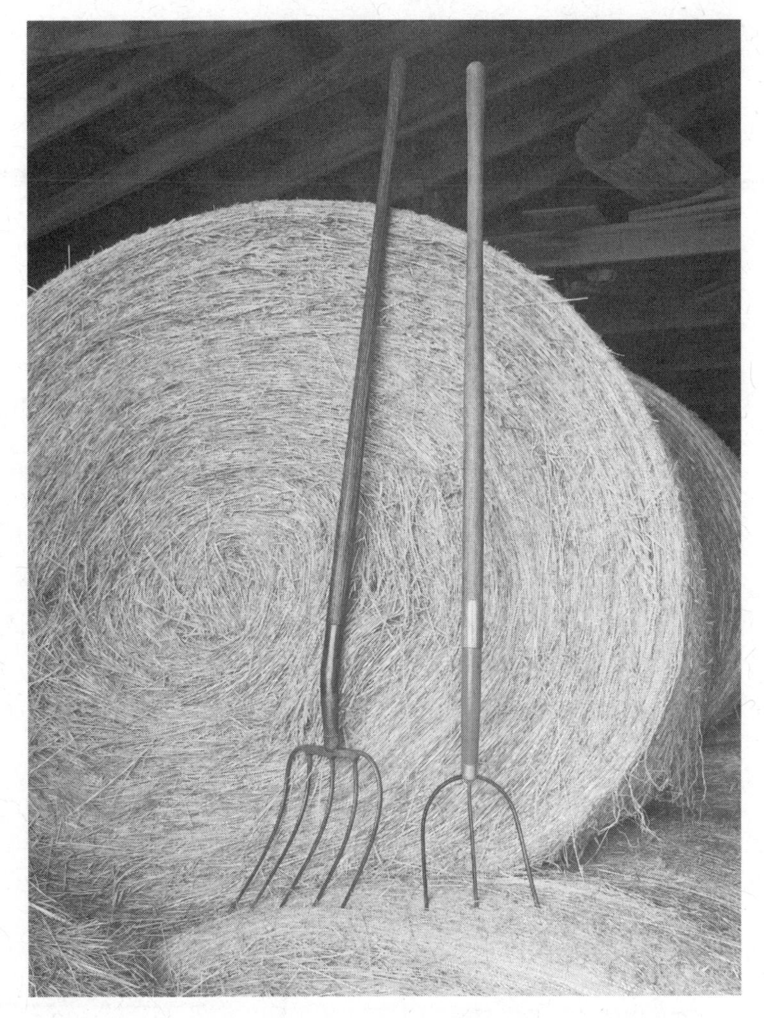

Fig. 8.4: *Hay forks come in many different styles. The pitching will go easier if the tines are relatively sharp and there are only three of them.* CREDIT: KAREN K. WILL

the hauling altogether and just make small haystacks (cocks) within a couple of rake's reach. If you choose this approach, you might cover their tops with tarps to avoid wasting more of the hard work than necessary. If you live where dews are heavy and/or fall frequently, you might want to create your stacks on horses, pallets, or other contrivances that will keep the stack's base from contacting the ground. Some sort of cover is recommended for small stacks under these conditions as well.

While the Sun Shines

Once you have the tools in hand, you're ready to go, assuming your hay meadow is ready to be cut. We generally try to make the first cutting just as the grasses are sending up their flower stalks or the clover is beginning to flower — it depends on the dominant species and which meadow it is. Contrary to ideal cutting conditions for mechanical hay making, the best time for cutting with a scythe is as the sun is rising. Dew-damp forage is more succulent and much easier to slice with a scythe than it is later in the day, after it dries out a bit. And, since the scythe tends to lay out the cut forage in a loose windrow, drying the dew isn't generally a problem, although folks that cut hay this way in the Northeast and other dewy areas will generally turn the windrows with a hay rake at least once during the curing process. Out in the Dakotas, much of Nebraska and parts of Kansas, this practice of mechanical hay "tedding" (stirring of the windrow to help chase moisture away) is practically unheard of; it is generally not required. When the dry haying winds of summer blow in those regions, even heavy windrows can dry down to less than 18 percent moisture in 24–30 hours.

One approach to mowing is to cut as much as you can in about an hour and leave the swaths to dry for a day or possibly two. When the humidity is high, come back later in the day to fluff and turn the swaths with the hay rake so that they will dry evenly. Normally, before the dew drops on the second evening after cutting, you will be able to rake the swaths into heaps and load the wagon. If you have a partner and if the swaths are thick enough, you can fork them directly onto a moving wagon — or you can rake two swaths together and then trade off forking and driving.

Unloading the wagon (or other transport device) is as easy as backing it into the barn and forking the load into a pile. Once you get the routine started — and assuming the weather cooperates — you can mow in the morning, then rake yesterday's cutting in the evening and haul it to the barn. After a couple of weeks of this, you'll have several thousand pounds of provender, and your belt will be a couple of notches closer to being entirely too big for you.

Summer Sunshine in December

We're often asked about the quality of handmade hay compared with machine made. Since we've never tested the protein levels on our handmade hay, we have no scientific data. What we do know is that our animals relish the stuff. The sheep will refuse their large round bales of still-green prairie hay as soon as the stack of handmade hay is uncovered. This phenomenon may have something to do with the yellow sweet clover that grows in some of our hay ground and the fact that the hay generally spends less time in the elements either drying or waiting to be hauled to the barn than the bales. In any event, even if you have a dozen ewes or a family cow to overwinter, you can make most, if not all, of the hay you need by whittling away at it over a month's time. And you can get yourself back in trim, just in time for summer.

When the Hay Cuts Short

Rather than fret about drought or other weather-related animal-feed issues, plan for some forage shortfall by judicious use of cover crops on the rest of your place and by double cropping some ground that you might have only thought of as your corn patch. The possibilities are endless; be sure to allow yourself the luxury of thinking outside the box and trying some wacky stuff that doesn't appear in the land-grant extension expert's handbook. What follows are some ideas for keeping your animals fed when the hay cuts short.

Plant a half acre of sorghum or milo — the seed is pretty inexpensive, and you can get a decent stand by broadcasting. Sorghums are quite tolerant of drought and will be graze-able by the time your cool-season pastures have gone away. Let the crop stand in the field and graze it off through the winter. You can get even more out of it if you control the grazed area with a portable electric fence so that the animals don't trample half of it looking for the most tasty of milo morsels, which of course, must be right over there!

Another crop that does fairly well with broadcast sowing is winter wheat. Seed in late fall for spring grazing or an early crop of hay. Follow it with a

Fig. 8.5: *Forages like alfalfa are a staple in some areas. These windrows are set up for baling, but they could also be gathered loose and the hay stacked or stored in the barn.* Credit: USDA-ARS, Doug Wilson

Feeding the Soil

Under the most ideal conditions, you'd want to feed the hay you make on the ground where you cut it. In practice, it may not be possible or practical to feed your animals on the ground where the hay came from. If you have to move your hay around, think about where you'd like to add some fertility, and feed the animals there.

Years ago, when I was raising Angus cattle, I fed all of the hay back to the cattle (with the exception of the first-calf heifers) on the ground where the hay was cut. Most of the hundreds of tons of hay I made never left the fields. I usually moved it to a fenced area at the fields' edges to keep the deer and cattle off it until it was time to feed. I also practiced an inexpensive feeding method by simply unrolling those huge hay bales in a windrow. The cattle would lineup along the row to feast. Sellers of hay feeders (and university advisers) cautioned me that most of the hay fed this way would go to waste because the animals would bed down in it, and the combination of compaction and hoof action would ruin the sod and soil. My solution was to keep the herd ever-so-slightly on the hungry side, so that the hay was gobbled up before anyone had time to think about taking a nap in it. Come spring, the meadow was in great shape; in fact, it grew taller, greener, and thicker — in stripes, where the hay had been fed.

When it came to the first-calf heifers, I kept them in a lovely pasture closer to the barn and fed them a combination of grass hay from some of my most productive meadows (almost six tons an acre in some years) and some alfalfa hay that I cut on a patch of leased ground. I moved the hay feeders around the area (about an acre) and ☞

Fig. 8.6: *Feeding hay out on the pasture or even the hay meadow is a means to help you move nutrients around your farm.* Credit: Karen K. Will

often wound, up with nothing but bare soil, piles of hay-turned-mulch, manure piles, and plenty of urine-soaked ground. I watched weeds grow lush in that area for a couple of years until one day, it dawned on me. Corn has a huge appetite for nitrogen, and, since the hay-feeding area was also a low area, there was usually plenty of soil moisture through the end of June. My thinking was that if I could pull some of those hay-supplied nutrients back into the corn, I could in turn ask the cattle to move it to other needy places around the farm.

So, I put up a temporary electric fence around that acre, borrowed a two-row, no-till drill from a good friend and planted it with a "grazing" corn. This particular corn was extra leafy — it put much less effort into making a pair of robust ears than it did in making plant biomass overall. And wow, did it come on strong! Dark green and thick. By early August, I was pretty much out of pasture for the cows, so I stripped off a couple of rows at a time, and, with portable electric fencing, gave the cattle access for part of the day. Then I sent them to the poorest-producing hay meadows, where they would graze the meager post-haying regrowth while depositing all those nutrients.

The cattle thrived on the corn and hay regrowth until autumn's rain and cooler temperatures brought on the fall pasture flush — had it not been for the corn, I'd have had to feed precious hay to get the animals through the summer.

Fig. 8.7: *Spread hay into a loose windrow when feeding in the winter — limit the quantity, and the animals will leave little waste to bed down in.*
CREDIT: OSCAR H. WILL III

mixture of rye and turnips. When the rye has been grazed off, let the turnips develop and then let your sheep or hogs at it. If you have access to a small no-till drill (check your local government farm service organization), let your ruminants graze their cool-season pastures hard before they go dormant in summer, then drill in some forage turnips. With a little luck, a few autumn rains will result in a nutritious crop that your sheep and cattle will relish.

If you're looking at a fairly hot summer, and the corn is ahead of schedule, or you planted short-season varieties, or just because, it might be time to procure some crabgrass seed to broadcast throughout your corn patch. Although it's an oft-loathed warm-season annual in northern lawns, crabgrasses are among the most productive forages in places like Oklahoma, where the pasture slump often arrives with triple-digit temperatures. Plenty of folks have great luck sowing crabgrass seed in the corn patch just before the last cultivation. The result is a nice, thick sward of good grazing by the time the pastures have turned off and the corn is picked. Some folks make hay with

More on Haying Your Way Into Shape

When I was starting out, I made hay for us and the neighbors — plus some to sell — using a combination of antique tools that included a 1956 John Deere 720 Diesel tractor, a 1950s vintage International Harvester model H with 9-foot semi-mounted sickle-bar mower, a 1960s model IH 5-bar rake, and a John Deere 14-T small square baler. I'm not sure about the baler's vintage, but it was 15 years old, at least. I handled most of the work myself, but when it came to baling, it took my two young daughters and me to get it done. I'd mow 30 to 40 acres at a time and usually rake it by noon the following day. Then we'd hook up the baler, and my then 9-year-old daughter would slowly navigate the tractor along the windrows, with baler and hay wagon in tow. My 11-year-old daughter and I would stack bales on the hay rack until it was full, then we'd disconnect it in the field and hook up another.

Once we had two racks full, we'd quit baling and hook both wagons to the pickup truck and head to the barn, unload the wagons, run the bales into the barn's loft on an old chain elevator, and stack them neatly. This went on for years. I can still remember how by the end of the season each year, I'd discard my belt — because it was seriously too long — and cinch up my jeans with baling twine. I'm sure such weight fluctuation would be frowned upon by pundits today — so be it.

As time went on, the girls grew to become lovely young women. They'd still help with the hay — and they'd usually have a small audience of fellows who could be pressed into wagon unloading and mow-stacking duties for pay and the chance to take lunch with the girls. And then one year, it happened. The guys we'd hired to handle the hay were somehow different. After unloading only ten bales from the first full rack, they bailed. I don't mean they secured the hay with twine; I mean they were suddenly nowhere to be found. Obviously bale handling was simply too tough for these young men.

So, I unloaded the wagons and put the hay into the barn myself (stopping only to drink and tighten my baling twine belt). Never before had I scrambled up and down the loft's ladder so many times in one day. The next year, I switched to large round bales. The result? My belt is no longer too large — even at the end of the season — even when I make a few tons with scythe and rake.

very tall-growing crabgrass varieties. Others hoard the hay they make and use additional forage options that can extend the grazing period overall or bridge the hot summer months when the cool-season options are nonexistent and even the warm-season forages are shutting down.

Some folks religiously plant cover crops once a primary crop is harvested. If you do this, don't forget that most of those crops make excellent forage — and some make great hay. It's true that those plantings will replenish the soil, but many can stand a single cutting or a light grazing with no ill effect. Convert everything you can to forage for your animals and you won't need to worry about how you'll feed those animals this year.

Chapter 9

Small Grains, Big Benefit

The backyard food garden brings ready images of labor converted to greens, fruits, beans, and corn — but why not small grains? There was a time when almost every farm and homestead out there made some provision for growing grains — and not only the larger grains like corn and beans, but also the small grains that continue to feed the world and that most folks in North America consume in large quantities every day. Is it because these staples have drifted into the realm of the highly processed and therefore are not for home growing?

Whatever the reason, it's true that dry beans, flint, flour and dent corns, along with a whole host of small grains are not among the top tier of food-garden favorites in North America. When was the last time you were involved in a conversation about the awesome flavor of a particular variety of corn, ground into meal, or which variety of wheat's flour was responsible for that oh-so-delicious homemade pasta?

No doubt, one practical reason for small grains falling out of favor is that there is quite a bit of labor involved with getting the crop out of the field and sufficiently processed to be able to make flour. And the making of flour requires milling — a process that we've been raised to believe requires vast stone or metal wheels powered by water or electricity to crush the grains and liberate that white starchy stuff we call flour. And we have a vague idea that it takes even more complicated machinery to separate the starchy stuff from the protein and the fiber — the germ and bran. And that's where we've gone

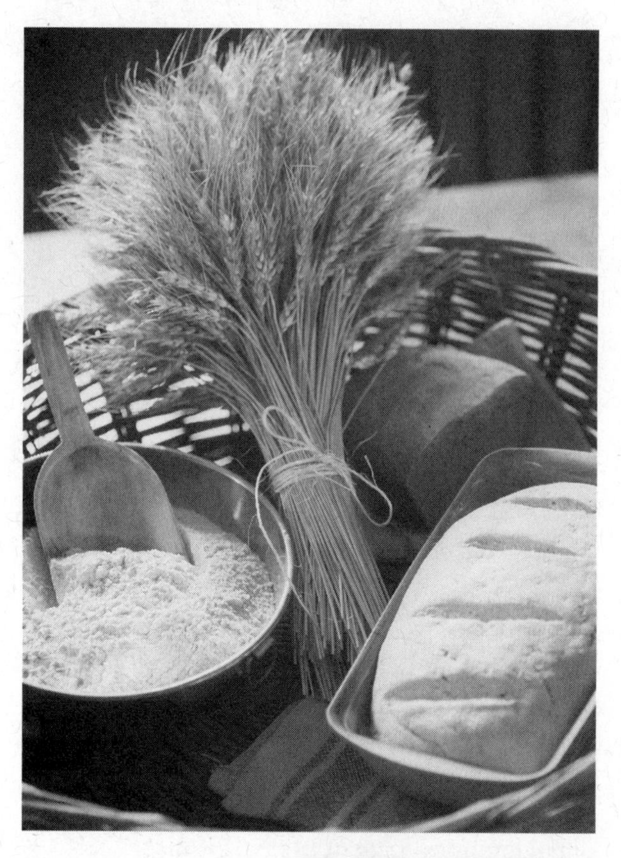

Fig. 9.1: *Grow the grain and grind the flour to make bread. We guarantee you will expend sufficient calories in the process to more than justify eating the bread with plenty of butter on it!* Credit: Karen K. Will

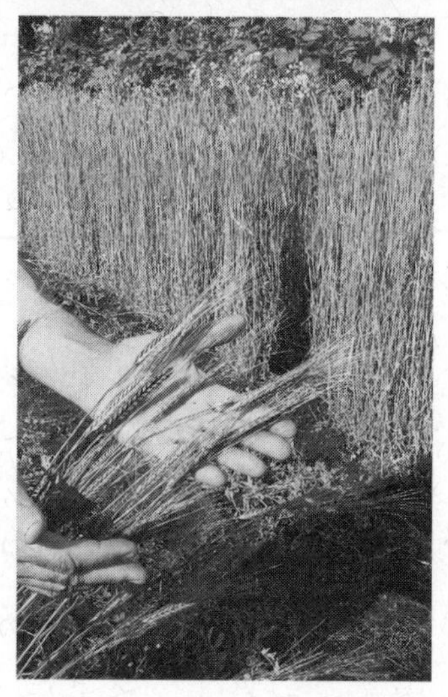

Fig. 9.2: *With its flat black heads, Black Einkorn is descended from some of the earliest domesticated wheat known. Not an extremely high yielder, this wheat may be prone to lodging when grown in warm humid areas.* Credit: Prairie Garden Seeds

wrong. Somehow, we've been convinced that the only flours worth eating are white and starchy.

The homestead is a perfect place to devote a little energy to the production of small grains. All it takes is a bit of fertility, some effort on your part, and minor cooperation on the part of Mother Nature. And come harvest time, if you don't have time to process the grain from the get-go, it will store quite nicely — so long as you keep insects, rodents, and birds at bay. Let's take a look at what's involved to make it happen.

Consider Your Needs

Before embarking on a full-scale small-grain production, try growing one to three small patches of your favorite grains (or some you are just curious about) as a means to dip your toes in the water. While those patches are growing, do some calculations to get a feel for how many pounds of whole wheat flour you consume in a year. Move from there to consider what fraction or multiple of an acre it would take, under average conditions, to produce that amount of grain. For example, if winter wheat is among the crops you want to grow, you might be able to achieve a yield in the realm of 40 bushels per acre, which roughly approximates one bushel per 1,000 square feet. So, if you devote a patch that's 20 feet by 50 feet to winter wheat, you should come close to making a bushel of threshed grain for your efforts. A bushel of wheat should weigh in the vicinity of 60 pounds — that's a lot of flour, when you think about it.

Putting in the Crop

Once you've decided on a wheat variety, you'll want to prepare your ground. You can do the initial prep with hogs, but you'll want to spend some time with a tiller, wheel cultivator, rake, etc., to get a fairly uniform seed bed. If you want to grow winter wheat, make sure to first check with local growers or the extension service (or online) to discover if the Hessian fly is an issue in your area; if it is, then find out what the earliest planting date is to avoid Hessian fly infestation. If you are in doubt, err on the later side for planting — but, with winter wheat, you'll need to give the crop sufficient time to germinate, sprout, and grow a bit before snow cover or freezing weather shuts it down. If you happened to have your peas, beans, or a cover crop of clover in a patch prior to the wheat, you won't have to worry too much about fertility. If you think fertility might be depleted, feel free to work some composted manure into the soil as you prepare the seedbed.

It's really not difficult to hand broadcast wheat seed in relatively small patches — you'll need three to four pounds of seed for a 40 x 50 patch (which would yield about two bushels total). As your patch grows larger, you might graduate to a broadcast seeder — either hand-cranked or a larger push or tow-behind model. If you have several acres to plant, it might be prudent to secure a small, solid-stand seeder or drill — you'll save on seed, and you'll be able to tell where you have and haven't seeded. Once the seed is on the ground, go ahead and rake it in, and then pack it. Again, if your scale is too large for the rake, fashion a drag — spike-tooth harrow, spring-tooth harrow, old bed spring, small snag of tree branches, you name it — and pull it over the patch to mix the seed into the seedbed. Plenty of folks leave it at this, but you can hedge your germination bet with a little pressure.

Fig. 9.3: *Hard red wheat, hard winter wheat, spelt, and Emmer (left to right, top to bottom) are all easy wheat-type small grains to grow in the garden and process by hand.*
CREDIT: TROY GRIEPENTROG

Fig. 9.4: *Winter wheat performs well when broadcast, as long as you follow seeding with a light cultivating and pack the seedbed — a cultipacker was used in this small field.* CREDIT: OSCAR H. WILL III

For the best germination, press the soil into good contact with the seed. If your patch is a small one, strap some 2 x 6 material to your feet and take a stroll. If you have access to an old push-type lawn roller, use it. If you have any easy-to-handle drum-sized cylinder around your place (plastic and metal barrels should be weighted), you can roll them back and forth to firm up the seedbed. Some folks drive a lawn tractor up and down the patch using a slightly different path with each pass. Others are lucky enough to have sufficient cultipacker wheels or drill press-wheels in their scrap piles to build small, easy to push or pull (by hand or with a lawn tractor) rollers that will get the job done. Crazy as all of this sounds, it will make a significant difference — and you can prove it to yourself by packing one half of your patch, observing and taking notes.

Once the soil is packed, all you really need to do is sit back, relax, and wait. If you are in an area that requires irrigation for any green life to thrive, you may need to sprinkle your wheat patch to get things moving.

What About Weeds?

Since folks generally sow wheat in a solid stand rather than in rows, cultivating really isn't an option. And you want to avoid walking on the crop to pull weeds by hand — which is neither fun nor efficient in this scenario. The best weed control you can offer wheat and other small grains planted in solid stands is good soil preparation — assuming you aren't into using poisonous herbicides. If the previous crop was a row crop, and you kept it well cultivated, you're off to a good start. If you've routinely used hogs to till and devour weed seedlings, you'll likely be fine. All you really need is for the wheat to get ahead of the weeds — winter wheat has the advantage of germinating at a time when plenty of other annual seeds are dormant, awaiting spring. Experiment a bit with rotations, rotation sequence, and cover crops to find a good working pattern for avoiding weeds in your wheat patch.

Another option for avoiding weeds is to plant your winter wheat in single rows so you can cultivate, or in multiple rows or thin strips that you can cultivate between. If you take this approach, you might plant six rows of wheat spaced 3 to 4 inches apart, and then leave a 2-foot path that you can protect with a heavy hay or straw mulch. This model will allow you to monitor the wheat and reach in from the outside to get after any major weed infestations. If you get way behind on the weeds, you can also mow them with your mower set on its highest setting to nip the weeds before they flower — most of them will perish when the frosts hit, and the wheat will have a head start in the spring. With a little luck, your efforts will pay off with relatively thick and uniform stands of wheat by the time the weeds are waking up — and they'll have a tough time of it under a canopy poised to turn into amber waves in just a few months.

Finally, if you do have difficulty with weeds in your small grains patches, consider using a densely growing clover in your rotation. To make this work for you, first go ahead and plant your wheat as you would normally, taking care not to stir the soil too deeply, thus exposing a new crop of weed seeds to the light of day. With a few good freezes over the winter, your winter wheat should come on pretty strong in the spring, outstripping the worst of the early weeds. Once the wheat is growing, but before the soil becomes tight and compacted from rains followed by dry weather, broadcast a low-growing

Sowing Wheat with a Precision Seeder

If you already own a walk-behind seeder that you use for planting row crops, you can put it to good use sowing wheat — you just want to put the rows as close together as you can — 3–4 inches works, if you can do it. If your wheat rows are spaced more than about 6 inches apart, you may experience reduced yields and added weed pressure in the patch. Using the planter will save seed overall and will press the soil into contact with the seed, but it may take a bit longer to accomplish, depending on the tools you have on hand for packing the soil. If your planter doesn't have a wheat plate (we doubt many will), just experiment until you find one that will pick up the wheat berries and space them a couple of inches apart in the row. Using the seeder might make more sense if you plant something in the vicinity of a quarter-acre or less, although a commercial-sized drop spreader (the kind your dad made you fertilize the lawn with when you were a youngster) might be adapted to sow a 36- or 40-inch swath in one pass — you'll want to rake, harrow, etc., and pack the soil, however. If you're inclined to modify a drop spreader for planting small grains, it might be time to find an old lawn roller or build a wheeled packing device to press the seed easily, evenly, and consistently.

clover (white Dutch comes to mind) over the wheat patch and water it in — or let Mother Nature do it for you. The clover will germinate and begin to grow under the "protection" of the wheat. By the time you harvest the wheat, the clover will be ready to take off, choking out any later-season weeds and eventually adding considerable nitrogen back to the soil. This method will require some experimentation and adjustment, but it often works. Be sure to follow the clover with something other than wheat (corn works); wheat can be planted again after the alternate crop has been harvested. We encourage you to consider a longer rotation series that includes other legumes if that fits your setup.

Harvest Time

Winter wheat generally ripens in early to late summer, depending on your elevation and latitude. The farther north you are, the later it will be. You'll know it's maturing as the verdant green turns silvery and then to various shades of amber, yellow, straw, or brown — there is plenty of variation in the color of mature wheat. At some point, you may notice that the seed stalks curve until the heads are aimed more-or-less downward. Pull a head or two, and rub it between your hands — if it is anywhere near ready, you should wind up with a small handful of loose wheat berries along with some chaff. Select a few berries and chew on them — crunchy means the wheat is ripe. This stage is ideal for combining wheat (mechanically harvesting and threshing with a modern combine), but you needn't worry so much about timing if you are going to be cutting and threshing your wheat by hand. Ideally, you want to harvest it before it is crunchy

Fig. 9.5: *This wheat is ready to cut although it might be just a tad too dry for ideal scything and hand threshing.*
CREDIT: OSCAR H. WILL III

and allow it to dry to the crunchy state before storing it.

When your wheat has turned mostly to amber (or red or whatever color your variety turns) and the berries are a bit on the soft or chewy side, you should feel free to begin your harvest. Use a scythe or garden sickle to harvest your wheat — if you happen to have a scythe with a cradle, so much the better. The cradle will collect the cut wheat stalks and keep them oriented head to tail for easy bundling. In any case, once you've cut sufficient stalks to make several bundles (perhaps a double handful of stalks at least 6 inches in diameter, measured at the stems), you want to gather it up, keeping the heads facing one direction and the stalks the other. Tie the bundles (sheaves) with a wheat stem or piece of twine. Shock several bundles by stacking them together (leaning against one another, teepee style) with the seed heads up. If you're worried about moisture, set a couple of bundles horizontally across the top. You could also simply haul bundles or loose wheat stems (with heads oriented together), and bring them into a well-ventilated and dry barn or mudroom where you can allow them to dry unmolested by rodents and birds until you can thresh out the berries easily with your hands; if they're crunchy, you're good to thresh and store. If they're still soft or chewy, you should allow the threshed wheat to dry before storing. Spread the berries on a tarp placed on the ground or a table to facilitate drying.

Getting at the Good Stuff

Finally, you're ready to separate the wheat from the chaff — literally. Humans carried out this operation for millennia before the advent of threshing machines or the modern combine, which harvests and threshes in a single pass. Luckily, ripe wheat shatters relatively easily, so all you need to do is rig up a threshing floor of some kind — and have at it. The threshing floor can be as simple as a cotton drop cloth, light canvas tarp or some other clean, aesthetically pleasing piece of material spread on a hard wooden or concrete floor. One of the simplest methods is to toss a few bundles into the center of the cloth, fold it in half or quarters, and then simply stomp on the enclosed bundles. You can crush the wheat in lots of other ways, so long as you use relatively light force, lest you crack the grain.

Some folks forego the folding and stomping and beat at the berry ends of the bundles with flails or sticks. Some use long-handled flails and work the bundles as they lay on the threshing floor (keeping the works on the cloth is helpful for collecting the grain), while others hold the bundles across their knees and whack at the heads with a short-handled flail or stick. Both methods may require a bit of vigor and rubbing to remove the seed completely from its hull. In time, you'll discover a method that works for you. If you happen to have a fairly powerful string trimmer on hand and don't mind the noise and smell, you can place bundles or wheat heads beaten from their stalks into a clean trash barrel and "stir" them vigorously with the machine until most of the wheat berries have been liberated — then it's time to winnow away the chaff.

Winnowing can be as high tech or as low tech as you desire. We suggest that you let Mother Nature do most of the work for you, as follows. Gather your threshed wheat in a large bowl or 5-gallon bucket, grab another vessel with as wide an opening as possible — like a galvanized wash tub about 24 inches in diameter. Take the works outdoors and into the wind. Set the wide-mouth vessel on the ground, grab a handful of threshed wheat, and drop it slowly from a standing height into the wide-mouth vessel. You might need to adjust your point of release depending on the wind conditions. As the grain drops, notice that the wind whisks away the chaff. Voila! You have clean wheat berries, or nearly clean, anyway. You might have to repeat the process a few times to remove 99 percent of the chaff; if the wind is fairly stiff, you might simply be able to slowly pour the wheat from the 5-gallon bucket into the wide-mouth vessel. Experiment with pouring from a few steps up on a stepladder — the longer the drop, the more chance for the chaff to catch the breeze.

Obviously, if you don't have natural wind, you can set up a winnowing station where you drop the wheat in front of a fan — we've also seen some disposable winnowers that folks cobbled up using a couple of cardboard boxes and a fan or two. Essentially, the wheat is slowly dropped into a tall box past paired openings through which a fan set up against the outside of the box can blow air in one side and out the other. If you get such a contraption to work satisfactorily, you could construct it using plywood or other, more permanent material for use every harvest season.

If the seed-dropping method doesn't appeal to you, you might try a method that was used by many of our ancestors on a relatively small scale. All you need is some wind and a winnowing tray. Find yourself a wide, shallow basket or build a broad, shallow lightweight box (use a cardboard box, if that's what you have) — add wheat and use the container to throw it up into the air in a fairly compact grouping. As the grain falls back into the winnowing tray, the natural (or human-made) breezes will carry some of the chaff away. Do it enough times with sufficient breeze and you'll wind up with wheat that's more than clean enough to cook with as is or to grind into various meals and flours.

Store the threshed and cleaned wheat berries (in plastic bags, glass, plastic jars, or other containers) in the freezer until you're ready to grind what you need for a few days. Besides preserving freshness, freezing will inhibit any moth or other insect eggs that may be trying to hitch a ride on your hard-earned grain. If you don't want to or cannot freeze your bounty, you might try sprinkling it with diatomaceous earth before closing it up tight in glass or plastic containers. If you happen to be a home brewer and can rig up a way to bathe (fumigate) your wheat (in a nearly sealed container) with the carbon dioxide escaping from the brew vessel's airlock, then you just might be able to kill any eggs, larvae, or pupae by depriving them of oxygen. If you go this route, you may as well allow the wheat to be in the CO_2 environment for a day or two. You'll still want to monitor the stored wheat carefully, and freeze as much as possible.

Use It Up

Wheat berries are incredibly versatile as a food source. Boiled or steamed, you can eat them much as you would pearl barley or rice — and you don't need to think so far ahead as to soak them overnight — although there's nothing wrong with soaking. Add two cups of winter wheat berries (be sure to sort for any stones that may have entered the scene during threshing) along with 6–7 cups of water to an appropriate-sized saucepan, bring to a boil, and then simmer for about an hour. Pour the works into a colander and rinse. You can serve this delight with butter and maple syrup, chill it and add it to a salad, or use it as a starch in place of pasta or rice. If you happen to own a rice-cooker machine, you can cook wheat berries in that. Add the recommended amount of berries and water for a "normal" batch of rice, and turn the machine on. It might take a bit

Fig. 9.6: *When you're sure that your cleaned grain is free of insect pests, it's safe to store small quantities in tightly sealed glass jars. This hard red wheat has been fine for a couple of months.*
CREDIT: KAREN K. WILL

longer than rice, and you might need to modify the ratio of grain to water some, but once you figure it out, you'll have fluffy, soft (not mushy) wheat to use in any way you can imagine.

If you get tired of eating wheat berries as just as a substitute for rice, you can "puff" them much as you would pop popcorn. The easiest method for puffing wheat is to place a small amount of oil in a heavy-bottomed saucepan, add a quarter cup or so of wheat berries, cover, and shake until they "pop." The popping sound won't be as loud as it is with popcorn, but the wheat will be rendered soft and crunchy — easy to chew. Some folks skip the oil and just keep the pan well agitated. Use the puffed wheat as you might use nuts — as snacks, in salads, crushed as a garnish, you name it. You can also use crushed or even more finely ground (run some through the blender, if you have one) puffed wheat as a breakfast cereal — just add some cream or whole milk, a bit of honey or maple syrup, and a dab of butter — wow! Try some crushed puffed wheat on your homemade vanilla ice cream with a dribble of maple syrup — double wow!

Piles of books have been written on using whole wheat flour for baking. What the research demonstrates is that never has anything been so controversial as the statement that it's impossible to make a palatable — even delicious — loaf of bread using 100% whole wheat flour. We are here to tell you otherwise. Quite contrary to popular milling and baking wisdom, it is *very much* possible to make delicious bread sans white flour! In fact, a very famous bakery brand — Pepperidge Farm — got its start when Margaret Rudkin (one of the farm's owners), created a 100 percent whole wheat flour bread that was delicious. She offered some to her physician, who was skeptical that bread could be made without white flour. And, as they say, the rest is history. An ironic footnote to this story is that way back in the 1930s, Mrs. Rudkin was interested in using whole grains, fresh fruits and vegetables, farm-fresh meats and fresh fish, along with less refined sugars like honey, maple syrup, and molasses. And her doctor — the catalyst for marketing the bread — agreed that it was a good diet, especially for children. The *Pepperidge Farm Cookbook* came out in 1963. Unfortunately (and inexplicably), Mrs. Rudkin didn't include her recipe for the bread that launched her business. And, sadly, today, the huge brand has strayed into the territory of highly processed and preservative-laden products.

Fresh whole wheat flour also makes excellent pancakes, muffins, and other quick breads. The first step in baking anything with whole wheat flour is to measure out your wheat berries and grind them into flour. If you don't have

Fig. 9.7: *Hand-grinding hard red wheat is a breeze compared with flint corn. With a setup like this, you can prepare all of your whole grain flours fresh in little more time than it takes to fill your canisters with commercial flour from a bag. You will be amazed at how much better it tastes fresh.*
CREDIT: KAREN K. WILL

a small grist mill, experiment rendering your wheat into flour using a blender. You can sift the result, if you'd like to get it really fine or remove the bran. Once you have a fine flour, it's time to make pancakes. Take a cup of the flour and mix it with 2 teaspoons of baking powder. Next, mix together 2 teaspoons of honey (adjust to suit your own taste), 1 cup of whole milk, 2 farm-fresh eggs, and a tablespoon of melted butter (substitute coconut oil, sour cream, cold-pressed sunflower oil, or cold-pressed peanut oil for variation). Finally, fold the works together and drop onto a hot, greased griddle. Serve these pancakes anyway you like — and don't be afraid to refrigerate leftovers — pack them in tomorrow's lunch pails to make some family members really happy.

Wheat berries can also be fed to your livestock, although only as a part of their diet. Chickens, hogs, cattle, or sheep will readily eat whole berries — you should probably limit the amount you feed the hogs because the berries swell in the gut. If you grind the berries first, your poultry will thank you and your livestock will digest quite a bit more of the good stuff contained in them. If you find yourself with a ton (literally) of extra shocked wheat, go ahead and feed entire bundles to the animals (remove any ties, unless you used wheat stems). Poultry will have fun threshing and eating; hogs and ruminants will enjoy munching seed heads, and they'll get a good dose of roughage at the same time. As always, feed grain as a treat rather than a mainstay to most of your animals.

Don't Chafe at the Chaff

One of the byproducts of growing small grains, threshing, and winnowing is that you wind up with a nice little pile of stems and chaff. As itchy as this material might be, it makes awesome animal bedding (especially the straw), excellent garden mulch, and perfect "brown" material for your compost pile. Even if you plan to use the stuff for mulch, consider that the hogs and chickens will enjoy having access to it beforehand, and they'll add plenty of value by working it into lovely compost that will benefit your garden even more. Clean, small-grain straw, especially oat straw, can also help keep your stock tanks and ponds clear by supporting and promoting the kinds of microbes that munch through excess organic matter and keep the aquatic nitrogen levels in check. Place the straw in mesh bags or tie tightly into bundles and float them on your ponds or tanks for best effect.

Beyond Wheat

Oats: Many other small grains are suitable for small-scale growing. Oats are a great choice if the homestead includes livestock and poultry. It's easy enough to plant a patch of oats as a cover crop, scythe and bundle the almost-ripe stalks and feed them whole over the course of the winter. For the table, oats are a bit more problematic because they tend to have a very tight hull that is difficult to remove. Even though bird damage can be a problem, hulless oats are preferred for homegrown table oats, unless you have a good full-service miller nearby who is willing to process small batches for you.

Rye: Rye can be successfully grown on a small scale. Though it's often sown as a cover crop — why not take some grain as well? Sow rye in the fall as soon as daytime temperatures consistently max out in the low 80°F range. If you get a good stand before winter sets in, you can lightly graze it and then again in the spring once it takes off. Famous for growing on poor soil, rye makes a great grain crop for areas in need of improvement. You can cut the grain close to the head by grabbing a bunch and carefully using a sickle — that way you can turn the remaining vegetation under, or otherwise crush it to build soil organic matter. For the table, process rye as you would wheat — most folks use it for baking — you might experiment with malting rye (sprouting, then drying or roasting and cracking or grinding) for brewing bread or for sprinkling on porridge, ice cream, salads — you name it. As a livestock feed, rye is less desirable than oats or wheat; animals don't seem to cherish it quite as much. Rye's value as a tenacious ground cover/green manure crop make it plenty valuable, but don't overlook its tasty grain to add some variety to your diet.

Barley: The diversity in barley types makes this small grain fun to grow. There are barleys with two rows of seeds, six rows of seed, spring-planted, fall-planted (winter barley), bearded and non-bearded. There are barleys that do well way up north and those that thrive in the heat. If you plan a barley experiment, research the various types and plant those that will be most likely to perform in your location.

Barley culture is similar to wheat's; if you plan to supply most of the water artificially, you might consider planting in wider rows (though broadcasting is fine, if that's the only means you have). Livestock tend to love barley; feed it in the bundle or threshed. If you really want to get the critters going, sprout the grain (it's OK to soak the heads still attached to the stems) before feeding — the animals' enthusiasm will be quite apparent. If you thresh the grain and malt it more traditionally, you can use the malt for beer making. You can sift out the hulls to use as a nice addition to baked goods and as a topping for ice cream or to make malted milk shakes. For direct use on the table, you'll want to remove the barley hulls. If you don't need barley pearls to be whole, you can try using a blender or your grist mill (set about as coarse as it will go) to break the hulls, then winnow the resulting hulls. Some folks take the processed grain, place it in a pan of water and float the released hulls away.

Fig. 9.8: *This rye was planted in closely spaced drills and will be harvested for long-cut rye straw for animal bedding instead of grain. In some areas, this straw is worth much more than the crop's grain value.*
Credit: Oscar H. Will III

There are many other grains and grain-like crops that you can grow specifically for the food and storage value of the ripe seeds — many turn out to be cover crops. Check out Gene

Logsdon's wonderful book, *Small-Scale Grain Raising* for all you ever wanted to know about raising, processing and using grains on a homestead-sized scale.

Whole Wheat Sourdough Bread

This is one of our favorite whole wheat bread recipes. We often use spelt flour, but more conventional whole wheat also works fine. Sourdough starter provides the leavening for this recipe; however, if you are at all worried about your sourdough's reliability, you can add ¼ teaspoon of fast-rise yeast to the dough. This bread has a lovely, dense crumb with a pleasant, tart flavor.

Fig. 9.9: *Sourdough whole grain bread.* CREDIT: KAREN K. WILL

2 cups sourdough starter
3¼ cups whole wheat or spelt flour
1¾ teaspoon fine sea salt
⅓ cup water

Place the starter, salt, and half the water in a large bowl and mix until the salt has dissolved. Using a rubber spatula or your hands, slowly mix in the flour. Add the other half of the water as the flour becomes harder to mix. Knead the dough for 10 to 15 minutes; the dough should be soft and easy to work.

Shape the dough into the desired shape and place in a well-buttered loaf pan; do not press it down into the pan. Using a very sharp knife, cut a few slits in the top of the loaf. Cover with a damp, lint-free cloth and set it in a warm spot to rise for 4 to 12 hours, depending on the temperature.

Bake at 350°F for 1 hour. Cool thoroughly before slicing. It will keep for one week without refrigeration (though we've never had a loaf last that long).

Section 4

Home(stead) As a Production Center

During the early years of our nation's settling, colonists, pilgrims, homesteaders, and pioneers all relied heavily on domestic skills to survive. Wood had to be chopped to fuel the hearth, vegetables had to be grown, and livestock had to be raised in order to have fresh food on the table. Food had to be preserved for leaner months, but also due to a lack of refrigeration. Cooking was an all-day, elaborate process that kept one or more women busy round the clock. In addition to the everyday chores of cooking, cleaning, milking, gardening, and preserving, our ancestors produced items of value in their homes — soap, fiber, clothes, furniture, pottery, and saleable crafts. For entertainment, they played musical instruments and read by firelight or candlelight. The home was a center of production and constant activity where "down time" was nonexistent, except for a few hours on Sunday.

Times changed, labor-saving devices were invented, and industrialization created inexpensive, mass-produced products that enabled the middle class to pursue leisure time and enjoy less-productive ways of entertaining themselves — television, video games, and shopping. We as a people, over several generations, lost our desire and ability to grow our own food, to cook meals from scratch, and to produce items necessary for everyday living.

History loves to repeat itself, and we're currently experiencing a revival of interest in self-sufficiency and returning to the home as a center of production rather than a site for mass consumption. By taking the reins of your life and doing more for yourself, in small but significant ways you'll detach from the chains of modern industry and the burdens that go along with it. One of our heroes, Shannon Hayes, coined this revival as "radical homemaking." Radical homemaking is the practice of turning one's home into a center of production, and relying on yourself, family, and community to produce most of what's necessary to live life.

The journey from consumer to producer takes some time, but beginning with cooking — the easiest task by far — is a good start. Learning to cook from scratch (in a kitchen that invites creativity and productivity), and perhaps even starting a home-based food business, will keep you firmly planted and connected to past generations. The lost art of domesticity doesn't have to remain lost. In this fast-paced world of instant gratification, take the time to develop skills of lasting value and consequence.

Chapter 10

Cooking from Scratch

Food is the key to our health, our state of mind, our well-being. Food is important, and it deserves a central place in all our lives. What we nourish our bodies with doesn't have to be an afterthought, it should be a forethought. When you think about food, reach deep down inside and tap into that instinct known as common sense. It will tell you: Eat real, whole foods. Eat in moderation. Grow, buy, or trade ingredients. Cook. Take a walk every day. Eat to live, don't live to eat. Follow traditional diets that have sustained people for centuries, rather than the modern, industrial diets that have been foisted upon us. Buzzwords like "heart healthy," "low cholesterol," and "fat free" come and go. Disregard such terms and start reading ingredient lists. They'll tell you everything you need to know to make decisions about what to buy and eat. It seems simple — and it is.

Our conversion to whole-food, dedicated from-scratch cooking began out of necessity. We had both lived rurally before, but nowhere like Osage County, Kansas. Twenty-five miles from the nearest proper city (Topeka), and situated eight miles equidistant between three one-street towns, we live in a county without a single stoplight. The ironic thing about this part of the country is that it's known as the "breadbasket of the world," yet most farmers around us can't even eat what they grow: genetically modified corn and soybeans. Restaurants and cafes get their food from the Sysco truck, rather than local farmers. A typical salad bar consists of bagged and shredded iceberg lettuce, processed cold cuts, and potato and pasta salads shoveled from a carton; choose from bottled Ranch, Thousand Island, or Italian dressings and top it off with Bac-O-Bits and croutons from a box, and there you have it. Entrees feature feedlot beef and pork and instant mashed potatoes; when

Just Read the Labels

When you quit processed food cold turkey, trips to the market become enlightening. As an example, instead of buying "non-dairy creamer" (whatever that is) for your coffee, you might choose half & half — which in its whole form is just half milk and half cream. Read the ingredients on several different brands, and you'll find that most contain things like corn syrup, carrageenan, dipotassium phosphate, sodium citrate, vitamin A palmitate, mono and diglycerides, and "color added." Why? These additives and preservatives are in there to mimic the real product (as "thickeners") or to prolong shelf life so the product can be shipped long distances and stay on the shelves, sometimes for more than a month. Keep looking until you find a somewhat local brand made the old-fashioned way. In our area, that brand is Anderson Erickson (from Iowa, one state over), and their expiration dates are usually a week away, sometimes less, from the date of purchase. However, the ingredients list reads: milk, cream. Just those two items. A handwritten sign states "no RBGH" (Recombinant Bovine Growth Hormone; handwritten, because labeling laws are being legislated). Amen. It can be done, but as a consumer, you must weigh the pros and cons: dairy products full of preservatives and additives of unknown origin that last a month in the refrigerator; or, whole, pure (and more delicious) dairy products that you need to use within a week. If you're cooking from scratch day in and day out, it's an easy choice. Seek out those manufacturers who are committed to real food, despite its shorter shelf life. Look for organic ingredients — you'll avoid genetically modified foods and their pesticide residues. Avoid anything that has a list that seems way too long and contains things you've never heard of, or that sounds like a bunch of elements off the periodic table — they are, in fact, someone's science experiment.

Fig. 10.5: Two ingredients — that's all you need! CREDIT: KAREN K. WILL

a genuine baked potato is offered, a basket with little plastic tubs of sour cream and "buttery spread" (margarine) accompanies it. Ask a server for some "real butter," and you're likely to get a blank stare; to most people, it's all the same thing. (Though, the restaurant experience isn't always this dire. The real, local, organic, farm-to-table food movement is growing by leaps and bounds — it just hasn't penetrated beyond the wall of Roundup Ready soybeans that surrounds us here.)

For practical reasons, we wanted to create healthy, delicious meals at home. We wanted to use whole, fresh ingredients, and we wanted to avoid the darlings of the modern industrial diet: refined vegetable oils, corn-fed meat, preservatives, fillers, and genetically modified foods — all of which are utterly devoid of nutrition, and harmful to your health. So, we picked up our knives and mixing bowls, and got serious about home cooking.

Taking charge of your food by cooking from scratch is an addicting, entirely enjoyable pursuit — no drudge work involved (drudgery is merely a state of mind). By avoiding the shortcuts of packaged, processed "foods," you'll not only save money, you'll hone your kitchen skills and evolve into a *very good* home cook, able to produce the tastiest victuals from the simplest ingredients — on the fly, mind you. Authentic ethnic cookbooks (e.g., French, Italian) are good inspiration because they always call for whole-food ingredients — no canola oil, no margarine, no "low-fat" or "fat-free" versions of anything — the way food was meant to be. We're always wondering why Europeans are so much healthier than Americans; the answer involves food. By eating a whole-foods diet you won't be filling your body with preservatives, pesticides, or artificial ingredients of any kind. Good food, prepared intentionally, is *true* health insurance.

Whole Foods

Here at Prairie Turnip Farm, we're blessed to have a freezer stocked with our own grassfed beef, pork, lamb, and broiler chickens. In the fridge are eggs from our flock of free-range chickens. We grow a kitchen garden, focusing on our favorite tomatoes, peppers, garlic, onions, chard, lettuce, spinach, and cucumbers. There's a bed for fresh herbs just outside the front door. Sourdough and vinegar crocks bubble on the countertop, and bread is baked at least once a week. These elements provide us with a good base from which to craft meals. There is always cooking going on in our kitchen. Producing our own food gives us an appreciation for what goes into it, and it makes us much more conscious of letting nothing go to waste.

For us, whole foods are good, honest foods that haven't been adulterated

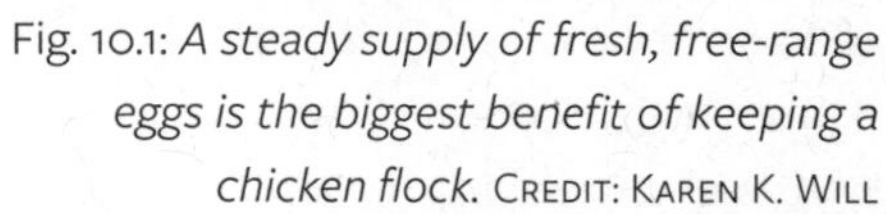

Fig. 10.1: *A steady supply of fresh, free-range eggs is the biggest benefit of keeping a chicken flock.* Credit: Karen K. Will

Fig. 10.2: *An herb bed, usually overrun with basil for pesto-making, is just outside the front door.* Credit: Karen K. Will

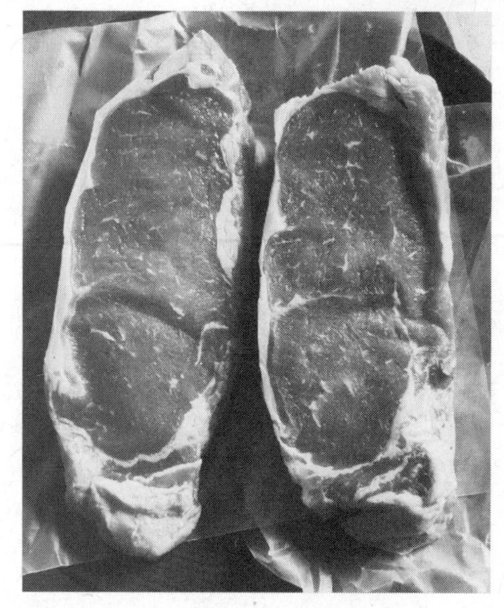

Fig. 10.3: *A freezer full of grassfed beef, pork, lamb, and chicken ensures you'll always have protein staples for homecooked meals.* Credit: Karen K. Will

in any way. We make exceptions for minimally processed ingredients like flour and sugar, but we limit our intake of them. Whole foods are full-fat, RBGH-free versions of all dairy products; whole grains; healthy fats like olive oil, coconut oil, home-rendered lard, butter, cold-processed sunflower oil, nuts, and avocado; real sweeteners like honey, maple syrup, and molasses; fresh fruits and vegetables; canned and frozen fruits and vegetables with no added ingredients; whole, free-range, pastured eggs; wild-caught fish; and grassfed, organic meat.

The Well-stocked Pantry

A well-stocked larder is a blessing, and a sort of insurance when you live far away from a 24-hour supermarket. "Ingredients" cost less per serving than convenience foods like frozen pizzas and TV dinners. When you only have ingredients to fall back on, you'll naturally become more creative in the kitchen, coming up with your own "recipes" and preparations. If you've got pasta, onions, olive oil, and some cheese, a meal is guaranteed. Throw in some fresh, dark leafy greens and a pound of ground pork, maybe even a carton of mascarpone cheese, and it's even better. Cooking with whole ingredients requires a little more imagination and planning on your part, but that effort will pay off in some of the best food you've ever cooked and eaten.

If you keep the following items stocked, you'll be able to make just about anything, whether it's from a printed recipe or a spontaneous meal of your own creation. Buy organic, if possible. Of course, individual tastes vary, so make substitutions where desired.

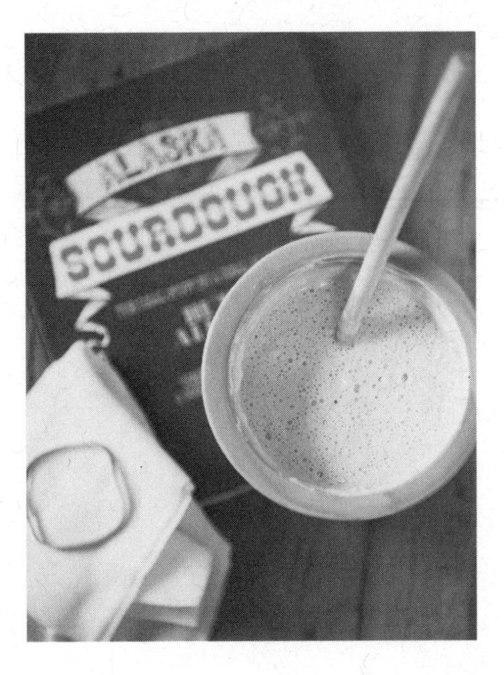

Fig. 10.4: *A sourdough crock bubbles on the countertop and provides the base for tangy breads, flapjacks, and muffins.* CREDIT: KAREN K. WILL

- Pantry:
    - Oats, oatmeal
    - Rice — brown and white
    - Pasta — white and whole wheat
    - Dried beans, peas, and lentils
    - Quinoa, couscous, and other grains
    - Cornmeal or dried corn for grinding
    - Nuts — pecans, walnuts, pinenuts, almonds
    - Flours — whole wheat, all-purpose, bread, spelt, rye
    - Potatoes (all-purpose Yukon Golds)
    - Onions (red, white, yellow)
    - Garlic
    - Chicken stock
    - Coconut milk (full-fat, not "lite")
    - Canned beans
    - Canned tomatoes — diced, crushed, pureed
    - Tomato paste
    - Pumpkin puree
    - Artichoke hearts
    - Roasted red peppers

- Olives
- Pickles
- Mayonnaise
- Dijon mustard
- Ketchup
- Natural peanut butter
- Brown sugar — light and dark
- Turbinado ("raw") sugar
- Sucanat (dehydrated sugar cane juice)
- Stevia
- Granulated sugar
- Local, raw honey
- Molasses
- Agave nectar
- Maple syrup
- Homemade jams and jellies
- Baking powder
- Baking soda
- Cornstarch
- Active dry yeast
- Quick rise yeast
- Cocoa powder (natural and Dutch processed)
- Chocolate chips (semisweet and bittersweet)
- Raisins and dried fruit (tart cherries, cranberries, blueberries)
- Dried coconut flakes
- Tea
- Coffee
- Sparkling mineral water
- Pure fruit juice

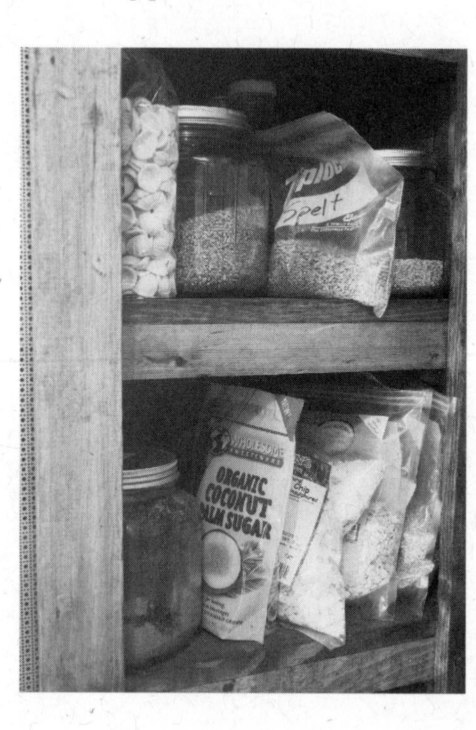

Fig. 10.6: *When you only have ingredients to fall back on, you'll naturally become more creative in the kitchen, coming up with your own recipes and preparations.* CREDIT: KAREN K. WILL

- Refrigerator:
  - Cheeses — cream, goat, mascarpone, ricotta, cheddar, mozzarella, Parmesan
  - Butter
  - Sour cream
  - Yogurt
  - Milk
  - Half & half
  - Eggs
  - Fresh vegetables (esp. dark leafy greens, peppers, carrots, broccoli)
  - Fresh herbs
  - Lemons and limes

- Freezer:
  - Grassfed ground beef, pork, lamb
  - Cuts of grassfed beef, pork, lamb
  - Bacon

   - Roasts for stewing and slow cooking
   - Whole and cut-up broiler chicken
   - Wild-caught seafood like salmon, cod, and shrimp
   - Lard
   - Frozen vegetables
   - Homemade tomato and pesto sauces
   - Homemade pastry shells
   - Homemade chicken/turkey/beef stock

- Spice Cabinet:
   - All varieties of dried herbs and spices
   - Kosher and sea salt
   - Tellicherry black peppercorns
   - Seeds — caraway, fennel, sesame, anise, poppy, pumpkin, sunflower
   - Extracts — vanilla, almond, lemon, maple
   - Oils — extra virgin olive, virgin coconut, peanut, virgin sunflower
   - Vinegars — red wine, white wine, apple cider, balsamic, malt
   - Hot sauce
   - Worcestershire sauce
   - Salad dressing mixes (pure herbs and spices)

Fig. 10.7: *Build up a righteous spice cabinet with all manner of seasonings, from common to exotic.* CREDIT: KAREN K. WILL

> Tip: Keep a dry-erase or chalk board near your pantry. When you use the next-to-last jar of something, write it down for reference on your next market day.

The Well-stocked Kitchen

At our house, we have a saying that goes "We can do all things through tools." With the proper tool in hand for a job, that job becomes a cinch. When it comes to cooking, it's not essential that your kitchen be stocked with every gadget and gizmo on the market, but having the right tool for the job will open doors for you that you never thought possible. For example, think about how to prepare a dish like scalloped potatoes from scratch ... With a mandoline, those ⅛-inch slices of potatoes are a snap. How about creamy tomato bisque soup? An immersion blender saves you from transferring soup in batches to a blender, and leaves you with only one pot to clean. Good kitchen tools clear up the vagaries of cooking, making preparation straightforward and hassle-free. They'll also allow you to produce consistent results with your food — a confidence-builder that will enable you to keep challenging yourself with more complex fare. Besides the usual kitchen staples, consider outfitting your kitchen with the following:

- Grain grinder — to grind cornmeal and flour
- Stand mixer (with dough hook, paddle, meat-grinding, and pasta-making attachments)
- Food processor
- Slow cooker — large enough for a whole chicken or chuck roast
- Food mill — for making tomato sauce
- Electric ice cream maker
- Mandoline
- High-quality, sharp knives — makes chopping a whole new experience
- Cast iron Dutch ovens (3.5-quart for bread making, 6-quart for stewing)
- Handheld electric mixer
- Immersion blender
- Instant-read thermometer — great for taking the temperature of water for yeast
- High-quality meat thermometer
- Pizza stone or cast-iron pizza pan

A Cooking Course

Whether you are a skilled cook or just starting out, you'll love working your way through the time-tested recipes that follow. Over the years, we've developed each one with a mind toward using staple ingredients found in our pantry. Each one helps build a cook's confidence and will convince you that you truly can make everything from scratch. Nothing is too complicated, and dishes tend to get better or change over time the more you make them and improvise.

Baking

When it comes to baking, the key is to do it often. The more often you bake, the better "hand" you'll develop, and, in turn, you'll become more of an intuitive baker. In the old days, recipes rarely stated exact measurements of ingredients like flour; instead, they simply called for "enough flour to make a stiff dough." Today, that statement strikes terror in most reader's hearts because we don't bake often enough to know what a "stiff dough" feels like. Get to know dough by working your way through these recipes, and many others.

No-knead Bread

Baking bread is one of the most useful skills you can develop as a cook, especially if you live in the country, miles from a decent bakery. When you start baking your own bread, you'll forever be "ruined" to store-bought bread or even bread from a good bakery, because it just won't taste as good as your own.

No-knead bread is the perfect entry to bread baking because it's insanely easy, and the results — an artisan-style boule with a chewy crumb and crisp crust — will build your confidence and encourage you to tackle more complex recipes and techniques such as kneaded breads, boiled breads, sourdough, and flatbreads.

To make no-knead bread, you'll need a couple special items: plastic dough scrapers, found online at specialty baking sites (try chefdepot.com or surlatable.com); and a 3.5-quart cast iron or ceramic Dutch oven (larger sizes will also work, though the loaf shape will end up being a little flat).

These recipes were adapted from Jim Lahey's no-knead method in his book, *My Bread,* (2009, W.W. Norton & Co.).

Wheat Bread

Each recipe makes one loaf

2¼ cups bread flour
¾ cup whole wheat flour
2 tablespoons ground flaxmeal
1¼ teaspoons salt
½ teaspoon active dry yeast
1½ cups cool water
coarse cornmeal and wheat bran for dusting

Rye Bread

2¼ cups bread flour
¾ cup rye flour
1 tablespoon caraway seeds
1 tablespoon fennel seeds
1¼ teaspoons salt
½ teaspoon active dry yeast
1½ cups cool water
additional rye flour for dusting

1. Combine all dry ingredients in a large mixing bowl and whisk together.
2. Add the water and stir with a large rubber spatula, folding the dough over itself until thoroughly mixed. You will have a slightly wet, sticky mass of dough; it will not form a ball. ☞

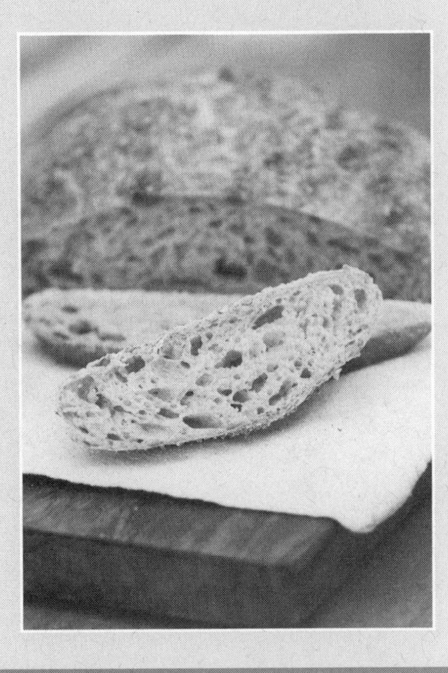

Fig. 10.8: Credit: Karen K. Will

3. Cover the bowl with plastic wrap and let sit at room temperature, out of direct sunlight, for 12–18 hours.
4. After 12–18 hours have passed, your dough should be dotted with bubbles and more than doubled in size. (It may also have a strong alcohol smell to it, but never mind that, it will burn off in the baking.) Dust a wood cutting board with bread flour, and, using plastic dough scrapers, scrape the dough loose from the sides of the bowl and turn out the dough onto the board in one piece. The dough will be loose and sticky, but do not add more flour. Dust the top lightly with flour and cover with a clean cotton or linen tea towel (terry cloth will stick and leave lint on the dough). Let the dough rise for another 1–2 hours.
5. About 30 minutes before the second rise is complete, place a 3.5-quart cast-iron pot (without lid) on the middle rack of the oven. Preheat the oven to 475°F.
6. Once the oven has reached 475°F, remove the pot using heavy-duty potholders (be very careful at this stage — the pot and oven are extremely hot). Sprinkle about 1/2 teaspoon of coarse cornmeal evenly over the bottom of the pot.*
7. Uncover the dough, and, using two plastic dough scrapers, shape the dough into a ball by folding it over onto itself a few times. With the scrapers, lift the dough carefully and let it fall into the preheated pot by slowly separating the scrapers. Dust the top of the dough with wheat bran.** Cover the pot and bake for 30 minutes.
8. After 30 minutes, remove the cover from the pot, rotate the pot, and continue baking for an additional 13–15 minutes until the loaf is browned, but not charred.
9. Remove the pot from the oven. With a sturdy wood or metal spatula, pry the loaf from the pot and transfer to a cooling rack. Allow the bread to cool for at least one hour, preferably longer, before slicing. The cooling time is necessary to complete the cooking process and shouldn't be overlooked!

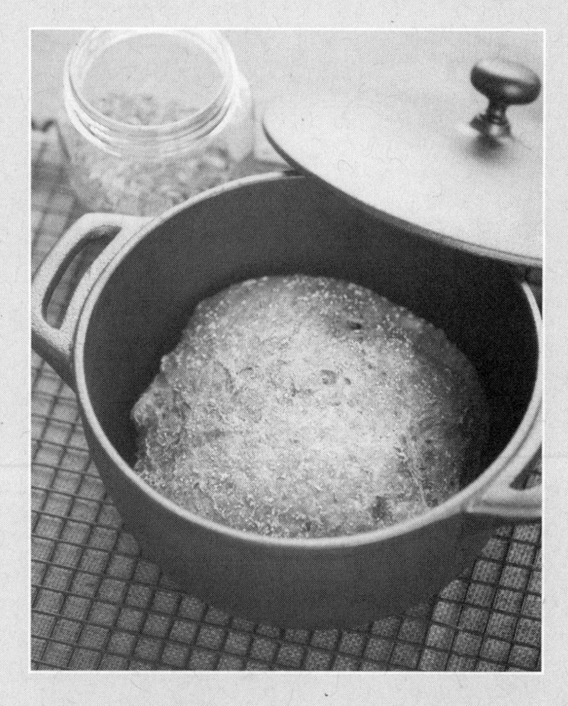

Fig. 10.9: CREDIT: KAREN K. WILL

*For rye bread, sprinkle rye flour on bottom of pot.
**For rye bread, dust the top with rye flour.

Baguette

This baguette is the core of our farm-based baking business, The Local Loaf (discussed in Chapter 12). Customers say it's addicting. It's unlike a traditional baguette in that it's soft and flavorful — you can easily tear off chunks from the loaf. It's perfect for dipping in olive oil or spreading with butter — or even on its own, thanks to the olive oil and salt we top it with. It's also good for sandwiches, and it makes a great base for bruschetta.

To make this bread, you'll need a baguette pan, which supports the unwieldy dough and turns out consistent, professional results. Matfer blued steel pans from France are perfect. You can find them online through specialty baking sites or Amazon.

Makes 2 baguettes

3 cups bread flour
½ teaspoon salt
¼ teaspoon active dry yeast
¾ teaspoon sugar
1½ cups cool water
extra virgin olive oil and flaked Kosher salt

1. Combine all dry ingredients (flour through sugar) in a large mixing bowl and whisk together.
2. Add the water and stir with a large rubber spatula until you have a thoroughly mixed, wet, sticky mass of dough.
3. Cover the bowl with plastic wrap and let sit at room temperature, out of direct sunlight, for 12–18 hours.
4. After 12–18 hours have passed, your dough should be dotted with bubbles and more than doubled in size. Dust a wood cutting board with bread flour, and, using your plastic dough scrapers, scrape the dough loose from the sides of the bowl and turn out the dough onto the board in one piece. Using your dough scrapers, fold the dough over and onto itself a few times to form a neat round of dough.
5. Pour a little olive oil into a bowl and brush the surface of the dough with oil. Sprinkle on a few pinches of Kosher salt, then cover loosely with a clean cotton or linen tea towel. Let the dough rise for another 1–2 hours.
6. About 30 minutes before the last rise is complete, preheat the oven to 475°F. ☞

Fig. 10.10: Credit: Karen K. Will

7. Once the oven has reached temperature, brush some olive oil in the baguette pan, coating all the surfaces. Uncover the bread and, using your dough scrapers, cut the dough circle in half. Separate the halves and, using the dough scrapers again, gradually work the dough to elongate each piece to about 12 inches. You may have to fold the ends under or stretch it a little with your hands to create an even baguette. Just don't overwork it or obsess about getting the perfect shape.
8. Dust your hands with flour and pick up each piece and transfer it to the baguette pan, stretching it a little as you move it. Brush olive oil over the top of each baguette and sprinkle a little more Kosher salt. Bake for 15 minutes.
9. Remove from oven and slide onto a cooling rack. Allow to cool for at least 1 hour before slicing.

Cornmeal Pizza Crust

Making your own pizza dough is another essential skill for a home cook. It's not complicated at all, and it can be done in about 15 minutes (plus rising time) with or without a stand mixer. Make a double batch and freeze the extras.

This pizza crust calls for some whole wheat flour, but feel free to experiment with the "ancient grain" flours like spelt and Kamut in this recipe, as well as sourdough — you really can't go wrong here. The addition of cornmeal to the pizza crust gives the dough a little heft, as well as an earthy taste and texture.

This dough can easily be made with your stand mixer, though the instructions are for making it by hand. To make it mechanically, simply combine the water, yeast, and honey in the mixing bowl, let sit, add the oil, then dump in all the dry ingredients and knead on low speed for 3–5 minutes.

Makes two large crusts, or four individual crusts

1 cup warm water (about 110°F)
1½ teaspoons active dry yeast
½ teaspoon honey
2 tablespoons extra virgin olive oil
1 cup whole wheat flour
1 cup all-purpose flour
1 cup bread flour
¼ cup coarse cornmeal
1 teaspoon salt

1. Whisk together the warm water, yeast, and honey in a large bowl. Set aside for 10 minutes to activate the yeast. When the surface appears foamy, add the olive oil and stir.
2. Whisk together the flours, cornmeal, and salt in a separate bowl. Add most of the dry ingredients to the liquid, blending with a wooden spoon until a ball forms. ☞

3. Turn out the dough onto a lightly floured work surface. Knead the dough for 5–10 minutes, working in the rest of the flour. When the dough is soft and pliable, place in an oiled bowl, and flip once to coat all sides. Cover with plastic wrap and set in a warm spot to rise until doubled or tripled in size, about 1 ½–2 hours.
4. Once the rise is complete, cut dough in half (or quarters), placing unused portions in plastic bags. Refrigerate if you plan to use within two days; otherwise, freeze.*
5. Preheat the oven to 450°F. Place a pizza stone or cast-iron pizza pan in the oven to preheat.
6. Roll out the dough into a thin round. Once your baking vessel is preheated, place the crust on it and layer on the toppings. Bake for 15 minutes or until browned and bubbling.

*To thaw, remove from freezer and allow to rest at room temperature for several hours before use. Do not thaw in the microwave (you'll kill the yeast prematurely).

Fig. 10.11: CREDIT: KAREN K. WILL

Pastry Crust

Having a fail-safe pastry dough in your arsenal of cooking skills is like money in the bank. This flaky pastry crust can be used to make not only sweet pies like pumpkin, but also savory pies like quiche. If you don't have access to pure lard, just use butter.

Tips: Make sure you have a large-enough work surface for rolling out the dough — ideally, a chilled surface like a granite countertop or a marble pastry board. If the kitchen is warm, you'll have to take extra measures such as freezing the dough in between all the steps and working really quickly and intentionally — no interruptions — or else the fats will melt and the dough will stick and tear. If you do experience such problems, try rolling it out on a piece of wax or parchment paper, then invert it onto the pie dish. Everybody's method for successfully making a pie crust is different — you just have to nail yours. ☞

Makes 2 pie shells

2¾ cups unbleached all-purpose flour
1¼ teaspoons salt
1 tablespoon raw sugar
1½ sticks (¾ cup) unsalted butter, very cold or frozen, cubed
¼ cup lard, very cold or frozen, cut up
½ cup ice water

1. Place the flour, salt, and sugar in the bowl of a food processor and pulse to combine. Add the butter and lard, and pulse until the mixture is crumbly, about 10 seconds. With the machine off, add the ice water and process just until the dough sticks together.
2. Dump the mixture onto a floured surface and knead for just a few seconds until the dough sticks together well. Divide into two equal parts and shape into balls.
3. Place each ball on a square of plastic wrap and squish down into disks. Wrap up the disks and refrigerate at least 1 hour before using; or, you can freeze for up to 1 month.

To prebake:

1. Preheat oven to 375°F.
2. On a lightly floured work surface, roll out one disk into a 14-inch round. Work quickly (so the fats do not start to melt), carefully, and intentionally — you want the dough to be even and ☞

Fig. 10.12: Credit: Karen K. Will

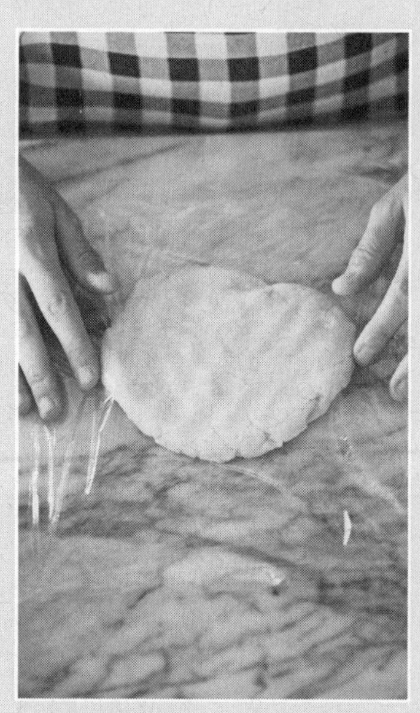

Fig. 10.13: Credit: Karen K. Will

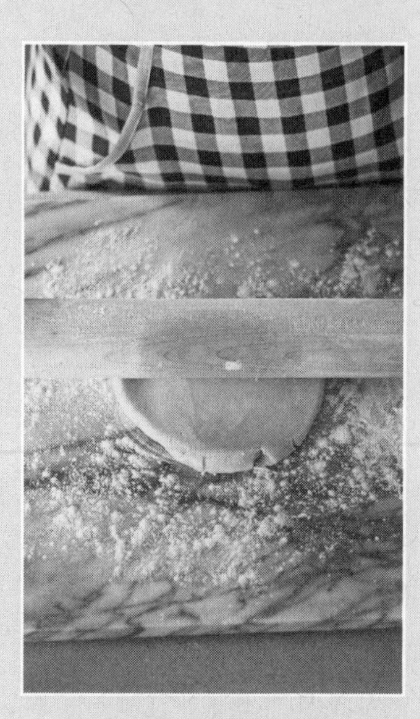

Fig. 10.14: Credit: Karen K. Will

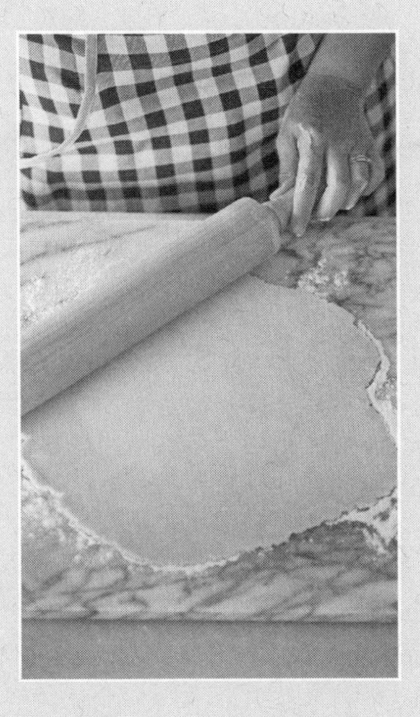

Fig. 10.15: CREDIT: KAREN K. WILL

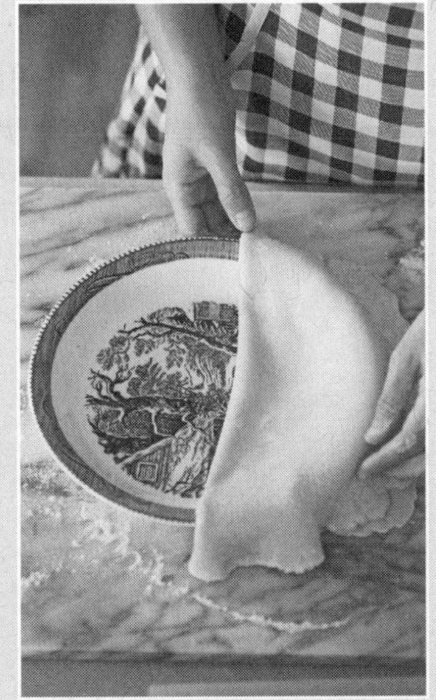

Fig. 10.16: CREDIT: KAREN K. WILL

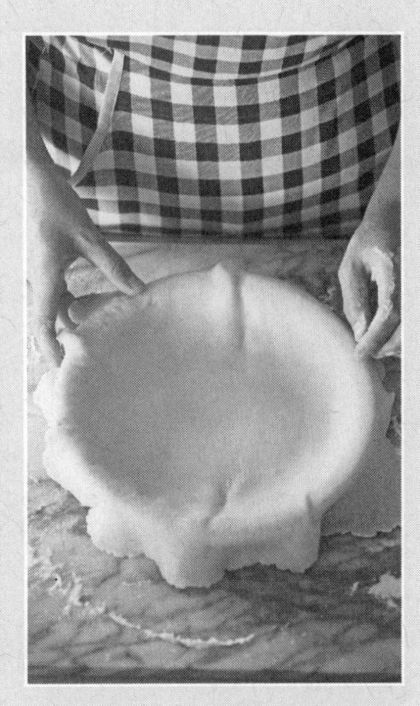

Fig. 10.17: CREDIT: KAREN K. WILL

not tear or fray. Center the crust on a 10-inch pie plate. Using a knife, trim the edges, leaving a 1/2-inch overhang. Fold the edges under, and press to seal. If you want to be fancy, crimp or flute the edges; for a more rustic pie, leave the edges simply folded under.

3. Using a fork, prick the bottom of the crust a few times. Place the pie plate in the freezer for 15 minutes before baking.
4. Cut a large circle of parchment paper (about 15 inches in diameter) and place it over the frozen pastry shell. Fill the shell with pie weights or dried beans.
5. Bake the pastry shell for 15 minutes; remove weights and parchment. Return to the oven and bake 15 minutes more, until golden brown. Cool completely on wire rack.

To bake:

Fill the prebaked shell with your filling and bake according to your recipe. If you are making a double-crusted pie and want a flaky top crust, brush whole cream over the top crust after filling the pie, and cut 3 or 4 slits to vent the steam. If it's a sweet pie (as opposed to savory), sprinkle on a little turbinado sugar as well. To prevent the crust from overbrowning, place an aluminum pie ring over the pie, or affix aluminum foil around the edges (tricky to do, but will work in a pinch).

Healthy Fat: Lard

In earlier times, lard was considered a very good, traditional source of fat. In America, cooks used it almost exclusively for pie crusts, frying, and myriad other uses, including soap making. But in 1953, American scientist Ancel Keys popularized the "lipid hypothesis" in his book *Eat Well and Stay Well,* which states that "there is a direct relationship between the amount of saturated fat and cholesterol in the diet and the incidence of coronary heart disease." This led to the belief that high-fat foods were "dangerous" and "unhealthy" — and the subsequent adoption of low-fat diets. The modern industrial diet with its emphasis on low-fat, fat-free, and "healthy" fats like canola oil and margarine, are simply the product of modern industry. The lipid hypothesis has many detractors, and solid research has placed its validity in question. What's important is that saturated fats from animal (and vegetable) sources provide needed energy in the diet. They provide essential building blocks for cell membranes, and they act as carriers of the fat-soluble vitamins A, D, E, and K. Fats from animal sources — lard, tallow, duck and goose fat — and vegetable sources — olives, coconut, flax — provide our bodies with highly beneficial fatty acids; they keep our bones healthy (by aiding calcium absorption), and they enhance the immune system. Engineered fats have none of these benefits.

Fig. 10.18: CREDIT: KAREN K. WILL

Lard, or pork fat, is about 40 percent saturated fat, 48 percent monounsaturated, and 12 percent polyunsaturated. The amount of omega-6 and omega-3 fatty acids varies in lard according to what the pigs have eaten, making fat from pastured or grassfed hogs the best choice. Lard is also a good source of vitamin D.

However, all lard is not healthy. The lard stocked on grocery store shelves has been harvested from "factory farmed" animals; it's been hydrogenated, deodorized, emulsified, and chemicalized. It's not anything you should be consuming.

Healthy lard, a source of beneficial saturated fat, comes from grassfed or pastured pigs, specifically, from the dense *leaf fat* that's deposited around a pig's kidneys. You can buy leaf fat at a butcher shop or small local meat processor (sometimes they give it away for free), or a local pig raiser. Once rendered, this type of lard has almost no pork flavor and can be used with excellent results in baking because the large fat crystals produce an exceptionally flaky crust.

Substitute lard for "shortening" in any recipe. Use lard in place of oil when frying, in pastry ☞

like pie crusts or cinnamon rolls, and for sautéing vegetables or roasting potatoes. You'll be delighted with the texture and flavor (and lack of pork flavor) that real lard provides.

Rendering lard is quite easy. There are two basic ways to do it; some say the water method produces a milder taste, but if done carefully and correctly, either way will produce a desirable product.

Lard

Dry Rendering

Chop the fat into chunks (if it hasn't already been run through a chopper). While it's still frozen, run it through a food processor with a metal blade or meat grinder to get the chunks even smaller. The finer the chunks, the less time it will have to spend in the oven.

1. Preheat the oven to 225°F.
2. Fill a large roasting pan with the chopped fat.
3. Roast slowly for 30 to 45 minutes until the fat has melted and protein particles (cracklings) are floating on top. The less time it spends in the oven, the better. Too much time in the heat will cook the protein particles which will impart a pork flavor to the fat.
4. Skim off the cracklings and set them aside for the chickens or dogs (though some people love to salt and eat these as a snack).
5. Place a mesh colander lined with a double layer of cheesecloth or butter muslin over a large bowl and pour the yellowish liquid fat through it. This will remove any remaining solid particles. Repeat using a clean cloth. Pour the liquid lard through a funnel into clean canning jars and allow to cool at room temperature for several hours.
6. Store in the refrigerator or freezer. Once completely cooled and solid, the lard will turn from yellow to snow white. It will keep for many months.

Water Method

This method can be done in a pot on the stove top or in a slow cooker.

1. Place the chopped fat in a medium-sized saucepan (the pan must be small enough to fit into your refrigerator). Cover with water and heat over low heat until the fat is melted and bits of protein are floating on top. (Tip: Add a halved potato to the pot to help soak up pork flavors.)
2. Turn off the heat and allow to cool. Place the pot in the refrigerator and chill overnight.
3. When the mixture is completely chilled, pop out the chunk of white lard setting on top of the water. The cracklings will have settled at the bottom in the water. Chop and store in glass jars or other airtight container.

Multigrain Cookies

Here's a cookie everyone will love. It's packed with whole grains and fiber, but deliciously sweet and filling. These morsels have a soft bite, yet crispy edges that have a deeply satisfying heft to them — you really can eat just one — in fact, we call them granola bars. They also freeze very well and can be eaten straight from the icebox.

Tip: When creaming the butter and sugar, don't rush it. Ideally you should beat for 3–5 minutes, until the mixture is no longer gritty. Creaming incorporates air into the batter, creating a nice rise in the cookie.

Makes 18 large cookies

1½ cups old-fashioned oats
1 cup whole wheat flour
½ teaspoon baking soda
½ teaspoon salt
1½ sticks (¾ cup) butter, softened
⅔ cup packed dark brown sugar
⅓ cup turbinado sugar
2 eggs
½ teaspoon vanilla extract
½ cup chopped pecans
½ cup unsweetened flaked or shredded coconut
1 cup bittersweet chocolate chips
¼ cup semisweet chocolate chips

1. Preheat the oven to 350°F. Adjust oven racks to mid-upper and mid-lower positions.
2. In a medium bowl, combine the oats, flour, baking soda, and salt. In a large bowl, cream the butter and sugars until smooth, 3–5 minutes. Beat in the eggs and vanilla until thoroughly combined. Stir in the dry ingredients into the wet ingredients, then stir in the pecans, coconut, and chocolate chips.
3. Use a ¼-cup measuring cup to drop cookies onto parchment- or Silpat-lined baking sheets.
4. Bake for 16 minutes until edges are browned, rotating trays from upper to lower racks and front to back, halfway through baking time. Let cool on baking sheets for 5 minutes before transferring to a wire rack to cool completely.

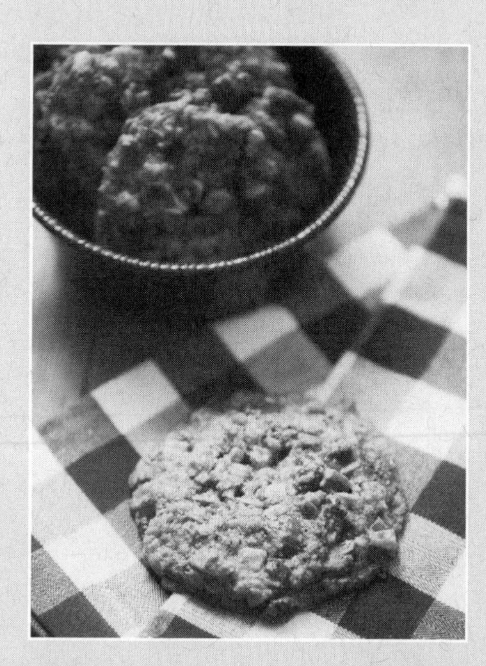

Fig. 10.19: Credit: Karen K. Will

Maple Pecan Scone

Scones are notoriously fussy, but this recipe makes them easy and does a good job of building confidence in your baking skills. Every scone recipe we've ever read carries the ubiquitous warning of not overworking the dough. This recipe actually calls for a light kneading of the dough so that it will come together thoroughly — instead of the usual crumbly mess that falls apart.

½ cup heavy cream (plus more for brushing)
1 large egg
1 teaspoon maple extract
3 tablespoons sugar, plus additional for sprinkling
2¼ cups all-purpose flour
½ teaspoon salt
1 tablespoon baking powder
6 tablespoons cold unsalted butter, cut into bits
½ cup finely chopped pecans
Maple icing (optional — recipe follows)

1. Preheat the oven to 400°F.
2. In a small bowl, whisk together the cream, egg, maple extract, and sugar until well combined. Set aside.
3. In a separate bowl, whisk together the flour, salt, and baking powder. Using a pastry blender, cut in the butter until the mixture resembles coarse meal. Stir in the chopped pecans.
4. Pour in the cream mixture. Using a fork, mix together until the mixture just forms a sticky but manageable dough. Turn the dough out onto a floured work surface and knead it gently for 30 seconds until it all sticks together. Working quickly (because you don't want the butter to get too soft), pat the dough into a ½-inch-thick circle, and cut out rounds with a 3-inch fluted* cookie or biscuit cutter.
5. Gather up the scraps and press together; cut more rounds. Place the scones on a parchment-lined baking sheet. Brush tops with cream and sprinkle with turbinado sugar (omit this step if making the icing).
6. Bake the scones in the middle of the oven for 15 to 18 minutes until golden.

(*A plain, round cutter would work, if that's all you have.)

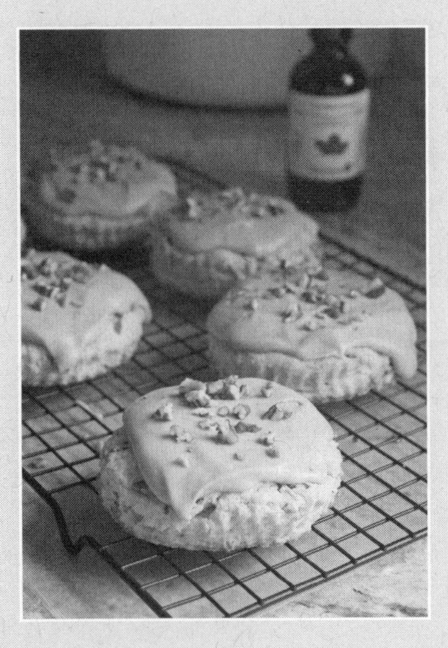

Fig. 10.20: Credit: Karen K. Will

Maple Icing

This thick, goopy frosting is literally icing on the cake. For holidays or special occasions, top the scones with it for the "wow factor."

8 ounces powdered sugar
2 tablespoons whole milk or half & half
2 tablespoons butter, melted
1 teaspoon strong coffee
Pinch of salt
1 teaspoon maple extract

In a medium bowl, combine all the icing ingredients and whisk until smooth. Pour over the scones (about 1–2 tablespoons per scone) and sprinkle with a few chopped pecans. Allow the scones to sit for 20 minutes before serving to allow the icing to set.

Real Food Substitutions

If you come across recipes for baked goods that call for vegetable oil, make the "real food" substitution that we do: equal parts sour cream and melted butter. For example, if the recipe calls for ½ cup vegetable oil, substitute ¼ cup sour cream plus ¼ cup melted butter. We've always been impressed with the results.

Must-Have Food Books

- *Real Food: What to Eat and Why* by Nina Planck
- *Nourishing Traditions: The Cookbook that Challenges Politically Correct Nutrition and the Diet Dictocrats* by Sally Fallon and Mary Enig
- *The Forgotten Skills of Cooking: The Time-Honored Ways are the Best — Over 700 Recipes Show You Why* by Darina Allen
- *Good Meat: The Complete Guide to Sourcing and Cooking Sustainable Meat* by Deborah Krasner
- *The Grassfed Gourmet Cookbook* by Shannon Hayes
- *Bob's Red Mill Cookbook: Whole & Healthy Grains for Every Meal of the Day* by Miriam Backes
- *My Bread: The Revolutionary No-Work, No-Knead Method* by Jim Lahey and Rick Flaste
- *Baking Illustrated* by America's Test Kitchen
- *New Vegetarian: Bold and Beautiful Recipes for Every Occasion* by Celia Brooks Brown
- *Serving Up the Harvest: Celebrating the Goodness of Fresh Vegetables: 175 Simple Recipes* by Andrea Chesman
- *Ratio: The Simple Codes Behind the Craft of Everyday Cooking* by Michael Ruhlman
- *The Flavor Bible: The Essential Guide to Culinary Creativity, Based on the Wisdom of America's Most Imaginative Chefs* by Karen Page and Andrew Dornenburg

Easy Meals

From-scratch cooking can be challenging after a stressful day working outside or in the office. When women ventured beyond the kitchen and into the workforce, the big food companies responded with their packaged, heat-and-serve convenience foods. If you consider cooking supper to be a reward for a hard day's work, you'll look forward to that creative time spent in the kitchen, honing your craft. The more you cook, coming up with something spontaneous to fix for dinner will be less of a challenge.

Creamy Roasted Red Pepper Soup

Soup from scratch is like a building project — you keep adding layers as you go. This is a very filling, satisfying soup that's easy to whip up. Serve it with some bread and a salad, and dinner is on the table. Soups invite creativity. Experiment with the method used here to devise your own soup recipes.

4 tablespoons butter
1 onion, chopped
1 red bell pepper, chopped
1 clove garlic, chopped
3 tablespoons all-purpose flour
2 cups chicken stock
2 large pieces jarred, fire-roasted red peppers
Chili powder
Cayenne
Seasoned salt
Pepper
½ cup heavy cream
½ cup milk
2 cups shredded cheese, your choice

1. In a large, deep saucepan, melt the butter over medium heat. Add the onion and pepper and sauté for 3 minutes. Add the garlic, sauté 1 more minute.
2. Whisk in the flour and cook for 3 minutes, stirring constantly.
3. Whisk in the chicken stock, making sure to break up any bits of flour.
4. Add the fire-roasted red peppers; season with chili powder, cayenne, seasoned salt, and pepper (use your best judgment for amounts). Bring to a boil, reduce heat and simmer for 20 minutes.
5. Whisk in the heavy cream and milk, cook another 5 minutes.
6. Using an immersion blender, puree in the pot.
7. Stir in 2 cups shredded cheese and cook another 2–3 minutes until thoroughly melted. Season with salt and pepper.

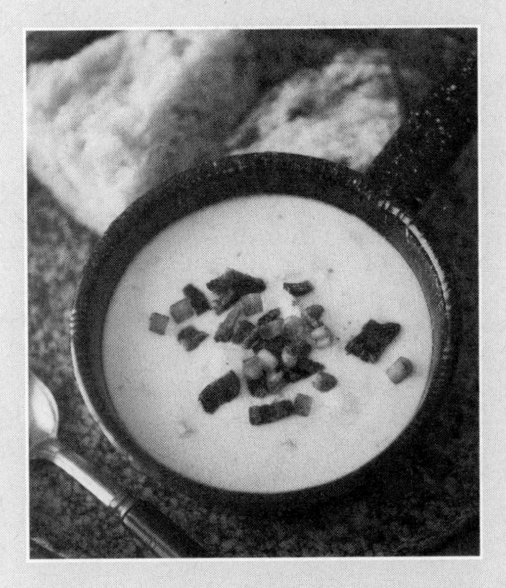

Fig. 10.21: CREDIT: KAREN K. WILL

Easy Weeknight Chili

This is probably the easiest chili you'll ever make because most of the ingredients are canned, making prep time only about 5 minutes. Use organic tomato products like Muir Glen, and organic beans, if available. Our favorite chili spice blend is "Chili Man" brand, which you can buy in some supermarkets or online.

Serves 4

1 pound ground beef, turkey, or pork
1 onion, chopped
1 cup water
1 (1 ounce) packet of chili spice blend
1 (15 ounce) can organic black or pinto beans, undrained
1 (15 ounce) can organic crushed tomatoes
1 (15 ounce) can organic fire-roasted diced tomatoes
1 (4 ounce) can fire-roasted diced green chiles or jalapenos

1. In a 6-quart Dutch oven, brown the ground meat with the onion.
2. Add the water and the package of spices and cook for 1 minute.
3. Add the beans, tomatoes, and diced green chiles, and bring to a boil. Reduce heat and simmer partially covered for 30 minutes, stirring occasionally.
4. To serve, garnish with sour cream, cilantro, shredded cheese, avocado, minced onions — whatever your heart desires.

Pasta with Greens and Mascarpone

This is a "go-to" pasta dish. It's creamy and slightly sweet from the mascarpone, but the bitter greens balance it out. It's ready and on the table in about 15 minutes. Substitute other vegetables like broccoli, asparagus, peppers, or tomatoes, as well as other cheeses like ricotta or chevre, to come up with variations. Serve with a toasted baguette.

Serves 4

8 ounces pasta (campanelle, bowtie, or penne are good ones)
3 tablespoons olive oil
1 onion, halved and sliced
2 cloves garlic, minced
¼ teaspoon crushed red pepper flakes (more, if you like spicy heat)
¼ cup chicken broth
1 large bunch Swiss chard, sliced into 1-inch ribbons
8 ounces mascarpone cheese
Lemon juice

1. Boil the pasta as you prepare the rest of the dish. Once cooked, leave it in the hot water until you're ready to use it, but you want to time it so it will be done about the same time as everything else.
2. Heat the olive oil in a large, deep skillet over ☞

medium heat. Add the onion and sauté for 5 minutes, until soft. Add the garlic and red pepper flakes and sauté for 1 minute.

3. Add the broth and greens, stir and cover. Cook until greens have wilted down, about 5 minutes, but not completely — you want a little life left in them — stirring occasionally.
4. Place the mascarpone cheese in a large serving bowl. Drain the pasta and immediately dump it into the bowl with the cheese. Add the greens mixture and toss everything, melting the cheese in the process.
5. Add a drizzle of olive oil and a squeeze of lemon juice. Serve with a pinch of sea salt and a sprinkling of pumpkin seeds or nuts of your choice.

Fig. 10.22:
CREDIT: KAREN K. WILL

Easy as Pie Quiche

If you've got pie shells in your freezer, this quiche can be fixed quite painlessly. A smorgasbord of ingredients can be substituted for the vegetables and meat, so it's a great, flexible recipe that you can adapt to your taste and whatever you've got on hand. If you've got the time and extra eggs, bake several pies and freeze them for future quick meals. To freeze, bake quiche as directed; cool completely, then cover with plastic wrap. To reheat, defrost quiche in refrigerator overnight or on the countertop for a few hours, then heat at 350°F for 20–25 minutes until hot through the center.

Makes 1 pie

3 eggs
1½ cups blanched broccoli, asparagus, or other vegetable
1½ cups cooked sausage, bacon, or other chopped meat
1 cup half & half, milk, or cream
½ cup grated cheese (your choice)
pinch of cayenne
pinch of salt
9" prebaked pie shell (see *Recipe: Pastry Crust,* above)

If you've got pie shells in your freezer, this quiche can be fixed quite painlessly. A smorgasbord of ingredients can be substituted for the vegetables and meat, so it's a great, flexible recipe that you can adapt to your taste and whatever you've got on hand. If you've got the time and extra eggs, bake several pies and freeze them for future quick meals. To freeze, bake quiche as directed; cool completely, then cover with plastic wrap. To reheat, defrost quiche in refrigerator overnight or on the countertop for a few hours, then heat at 350°F for 20–25 minutes until hot through the center.

Chicken Pot Pie

Pot pies are good any time of year, but especially during the cold months — when comfort food calls our name. It's not as complicated as you might think. If you don't have leftover chicken in your refrigerator, this same recipe can be made with ground beef, beef broth, and chili powder (in place of the poultry seasoning). Further, substitute bell pepper for the celery, and corn for the peas. For a little flavor and crunch, substitute a couple tablespoons of cornmeal for the same amount of flour in the crust. Now we're improvising!

Serves 6

Crust:
1½ cups all-purpose flour
½ teaspoon salt
½ cup butter (1 stick) or lard (or a combination), cut into cubes
5 tablespoons ice water

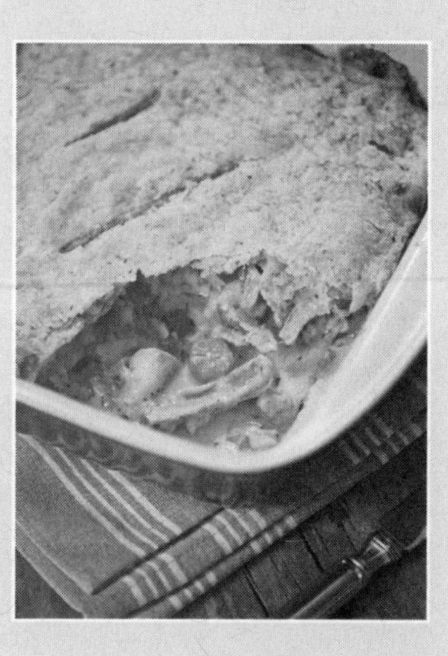

Fig. 10.23: CREDIT: KAREN K. WILL

Filling:
2 tablespoons butter
1 cup chopped onion
1 cup sliced celery or fennel
1 cup chopped carrots
¼ cup all-purpose flour
1 teaspoon poultry seasoning
½ teaspoon salt
½ teaspoon pepper
1 cup chicken broth
¾ cup whole milk or half & half
2½ cups cooked chicken meat, chopped
1 cup frozen peas
¼ cup fresh parsley

1. Stir together the flour and salt in a medium bowl. Using a pastry blender, cut in the butter until the mixture resembles coarse meal.
2. Sprinkle 1 tablespoon ice water over the mixture and toss with a fork. Sprinkle additional water, 1 tablespoon at a time, over the mixture until it's moistened and comes together in a ball. Refrigerate the pastry until the filling is ready.
3. Preheat the oven to 400°F.
4. In a large saucepan over medium-low heat, melt the butter. Add the onion, celery, and carrots, and sauté for 5 to 7 minutes, until tender.
5. Stir in the flour, poultry seasoning, salt, and pepper, and cook for 3 minutes. Whisk in the broth and milk. Cook until thick and bubbly. Stir in the chicken, peas, and parsley.
6. Butter a 2-quart casserole dish or an 8 x 8 x 2 baking dish. ☞

7. Turn the pastry dough out onto a floured surface and roll out the dough 1 inch larger than your dish.
8. Pour the hot filling into the dish and place the pastry on top. Use a fork to crimp the edges, flute them with your fingers, or just tuck them under for a rustic-style crust. Use a sharp knife to cut slits in the top to allow steam to escape.
9. Bake for 30–35 minutes, until golden brown and bubbling. Let cool for 20 minutes before serving.

Dairy

Crafting your own dairy products takes home cooking to the next level. If you have access to farm-fresh (or raw) milk and cream, you'll marvel at the flavors and consistency of homemade cheese and ice cream. If you must rely on store-bought dairy products to make your cheese and ice cream, avoid products that have been ultra-pasteurized, a process that strips out all the flavor.

Easy Cheese: Mozzarella

If you've never tried making your own cheese, get a good home cheese making book, like *Home Cheese Making* by Ricki Carroll (2002, Storey Publishing), and experiment with the recipes. The recipe given here is adapted from her book. Cheese making is addictive, just like the finished product. Try your homemade mozzarella on pizza or eat it Caprese-style, with sliced tomatoes and fresh basil.

Makes ¾ to 1 pound

1½ teaspoons citric acid, dissolved in ½ cup cool water
1 gallon pasteurized whole milk (not ultra-pasteurized)
¼ teaspoon liquid rennet (available online through cheesemaker's suppliers), diluted in ¼ cup cool, unchlorinated water
1 teaspoon cheese salt (flaked, non-iodized Kosher salt)

1. Pour the milk into a stockpot. Take a temperature reading of the milk; when it warms to 55°F, stir in the citric acid.
2. Heat the milk over medium-low heat until it reaches 90°F. The milk will begin to curdle.
3. Stir in the diluted rennet, using an up-and-down motion with the spoon. Continue to heat the milk until it reaches 100–105°F. Turn off the heat. The curds should be pulling away from the sides of the pot. After 3 to 5 minutes, they will be ready to scoop out.
4. The curds will start to look like shiny, thick yogurt, and the whey will turn from milky white to clear. If it's still milky, wait a few more minutes. ☞

5. Using a slotted spoon, scoop out the curds and transfer to a 2-quart microwavable bowl. Press down on the curds gently with your hands, and pour off as much whey as possible. Reserve the whey.
6. Microwave the curds on high for 1 minute. Drain off all the whey. Gently fold the cheese over and over (like kneading bread) with your hands or a large rubber spatula.
7. Microwave two more times for 35 seconds each; add salt to taste after the second time. After each heating, knead again to distribute the heat.
8. Knead the cheese quickly until it is smooth and elastic. When it stretches like taffy, it's done. If the curds break instead of stretch, they are too cool and need to be reheated.
9. Once the cheese is smooth and shiny, roll it into small balls.
10. Eat it while warm or place it in a bowl of ice water for 30 minutes to lower the inside temperature rapidly. Cover with plastic wrap or place in an airtight container and use within 2 days.

Brown Bread Ice Cream

The phrase "to die for" is used a lot when it comes to food, but we've reserved its use for this ice cream ... it really is *to die for!* It's a custard-style base with caramelized bits of wheat bread threaded throughout — no one can ever guess those bits are bread. Brown bread ice cream is traditional in Britain, where farm women often use fresh goat milk. We usually make it that way as well, but rest assured, once cooked, no "goaty" flavor comes through in the ice cream. Cow's milk works equally well.

Makes about 2 pints

2 ounces whole wheat bread
¼ cup packed light brown sugar
2 tablespoons butter
1 cup whole milk or goat milk
1½ teaspoons vanilla bean paste or ½ vanilla bean, cut open and seeds scraped
4 egg yolks
½ cup granulated sugar
1 teaspoon cornstarch
1 cup heavy cream

1. Preheat the oven to 350°F. Line a baking sheet with parchment paper and set aside.
2. Crumb the bread in a food processor and set aside.
3. In a large skillet, melt the butter and brown sugar over low heat. Stir in the bread crumbs and coat well.
4. Turn the bread crumbs onto the prepared baking sheet and spread them out. Bake for 15 minutes, turning several times throughout the cooking time, until the crumbs are browned and crunchy. Set aside to cool. ☞

5. Into a large saucepan, pour the milk and vanilla paste. Heat over medium heat, stirring constantly, until almost boiling, then remove immediately from heat. Set aside for 15 minutes to cool and allow the vanilla flavor to infuse. (If using a vanilla bean, remove the pod after infusing.)
6. In a large mixing bowl, beat the egg yolks, sugar, and cornstarch until the mixture is thick and light in color. Add a ladleful of the warmed milk mixture and whisk. Continue adding the milk and whisking after each addition.
7. Pour the mixture back into the saucepan and cook over low heat, stirring constantly. The custard is done when it has thickened and coats the back of a spoon. (You should be able to remove the spoon from the custard, run your finger across the back of it, and the line doesn't close up.) Be careful not to overcook, or the mixture will curdle.
8. Remove from heat, stir in the heavy cream, then set the pan in a bowl of ice water to begin cooling immediately; let sit for about an hour.
9. When cool, pour the cooled mixture into an electric ice cream machine. Churn for 20 minutes.
10. Meanwhile, run the cooled breadcrumbs through your fingers to break them up into small pieces. After 20 minutes of churning, add the caramelized breadcrumbs and churn for another 5 minutes. Immediately transfer the ice cream to a container and freeze for a few hours before serving.

When ready to serve, take the ice cream out of the freezer and let sit at room temperature for about 10–15 minutes to soften.

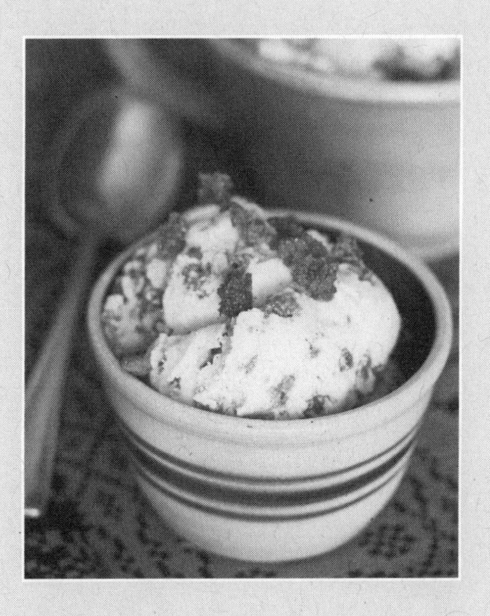

Fig. 10.24: Credit: Karen K. Will

Sauces & Condiments

Reading the labels on commercially made condiments, it's hard to imagine something edible was produced with all those random things. Tie on an apron and do a little experimenting. Making your own condiments and sauces is fun, and a real eye opener. Keep in mind that homemade versions don't have the shelf life of store-bought because they don't contain preservatives, so plan to use them up over the course of a week or two.

Mayonnaise

According to Julia Child in *Mastering the Art of French Cooking*, "mayonnaise ... is a process of forcing egg yolks to absorb a fatty substance, oil in this case, and to hold it in thick and creamy suspension." Making mayonnaise by hand is a beautiful thing, though the hard part for us was discovering the right type of oil that would duplicate the taste of our favorite store brand — Hellmann's — which uses soybean oil. After much trial and error (mostly error), we've determined the key is light-tasting olive oil, a healthy fat that doesn't impart the bitterness of extra virgin olive oil.

Fig. 10.25: CREDIT: KAREN K. WILL

Makes about 1¼ cups

3 egg yolks
2 teaspoons white wine vinegar
1 teaspoon Dijon mustard
¼ teaspoon table salt
1¼ cups olive oil ("extra light" — not extra virgin)
Ground white pepper (optional)

1. Place the egg yolks, vinegar, mustard, and salt in a large mixing bowl. Using a handheld electric mixer on the highest speed, beat until the mixture is frothy.
2. Beginning with a drop or two, add the oil and beat until the emulsion process begins. Beat until you cannot see oil any longer, but a thickening sauce. Keep adding the oil, increasing the amount with each addition. Remain patient and do not rush this step; this process will take about 10–15 minutes for the egg yolks to fully absorb the oil.
3. If the mayonnaise is too thick for your liking, beat in 1 to 2 tablespoons of boiling water to thin and smooth the sauce. Season to taste with white pepper and additional salt. Store in an airtight container for up to 3 days.

Tip: You can make all kinds of quick, delicious sauces, dips, and condiments (for grilled fish, vegetables, or whatever) just by mixing together equal parts mayonnaise and sour cream, then adding fresh or dried herbs, fresh horseradish, hot sauce, mustard, garlic, or whatever you crave.

Barbecue Sauce

Homemade barbecue sauce is a real treat for your family and friends. We make it whenever we smoke a pork butt or ribs; its complex, rich, sweet flavor marries well with the meat. It's also delicious on pizza, topped with onions, cheddar, and cilantro. It takes just 20 minutes from start to finish (including 15 minutes cook time), and the result is truly fantastic.

Makes about 1½ cups

¾ cup tomato puree (sold as heavy puree)
¼ cup pure maple syrup
3 tablespoons molasses
3 tablespoons organic ketchup
3 tablespoons malt vinegar
3 tablespoons Worcestershire sauce
1 tablespoon Dijon mustard
1 teaspoon garlic powder
½ teaspoon smoked paprika
Sea salt and black pepper

1. Combine all the ingredients in a saucepan. Heat over medium heat until boiling, reduce heat and simmer gently for 15 minutes, until reduced and thickened slightly.
2. Season with additional salt and pepper, to taste.
3. Store in an airtight glass jar or bottle for up to 2 weeks in the refrigerator.

A Bit of Homespun Fun

Hank and I have come to the realization that cooking at home is much better, and more fun than going out. But that doesn't mean we don't crave the restaurant experience from time to time, so we've come up with ways to create it without ever leaving the farm.

A couple times a year, we order live lobsters from a lobster pound near Hank's parents' home in Portsmouth, N.H. The lobsters are delivered overnight, and we usually schedule a Friday delivery to kick off the weekend. I'll make side dishes of potato salad or coleslaw, and fresh sourdough bread. We boil the lobsters whole, along with some corn on the cob. Then we set a big bowl on the dinner table for the shells and commence cracking and ferreting out all the delicious, sweet meat from the body. There's always a healthy chuckle when we suck out the legs or come across something interesting — like the female's eggs. It can take quite a while to pick the lobster clean, but we love this time, sipping wine and relaxing with our luxurious meal.

Though having lobsters delivered overnight isn't cheap or sustainable, we're supporting a small, independent American fishery, so we feel good about that. Figure out what is fun and affordable for your family — homemade pizza or pasta night, perhaps — that makes cooking at home an adventure and gets everyone involved.

Fresh Heirloom Tomato Ketchup

Fresh ketchup cannot be compared to the bright red, cloying, corn-syrup-laced material that comes from a squeeze bottle. This recipe is quite simple — it doesn't require any peeling or coring of the tomatoes — and it's worth the time spent in the kitchen. Make enough to last throughout the year.

Makes 2–3 cups

3 pounds ripe heirloom tomatoes, chopped
1 medium onion, minced
3 garlic cloves, crushed
1 tablespoon black peppercorns
¼ teaspoon dry mustard
½ teaspoon ground allspice
2 whole cloves
2 teaspoons celery seeds
1 inch of a cinnamon stick, broken
1 teaspoon smoked or sweet paprika
¼ teaspoon cayenne pepper
⅓ cup brown sugar
⅓ cup apple cider vinegar
juice of ½ lemon
1 teaspoon sea salt

1. Place the first 11 ingredients (through the cayenne) into a stockpot and bring to a simmer. Cook gently for about 40 minutes, stirring regularly until about ⅓ of the juices have evaporated. Let sit for at least 30 minutes to cool.
2. Puree the tomato mixture in the pan using an immersion blender, or use a blender and process at the highest speed for 1 minute. Pulse it to start — so you don't get splashed.
3. Run the mixture through a food mill using the finest-mesh screen (or use a chinoise or fine sieve) and return to a clean saucepan.
4. Heat to simmering, add the sugar, vinegar, lemon juice, and salt; adjust spices to taste. Simmer for about 1 hour, stirring frequently, to thicken the ketchup. Allow to cool to room temperature.
5. Freeze in freezer jars, or refrigerate for up to 3 weeks.

To use frozen ketchup, thaw and simmer for 15–20 minutes to allow the extra water to evaporate.

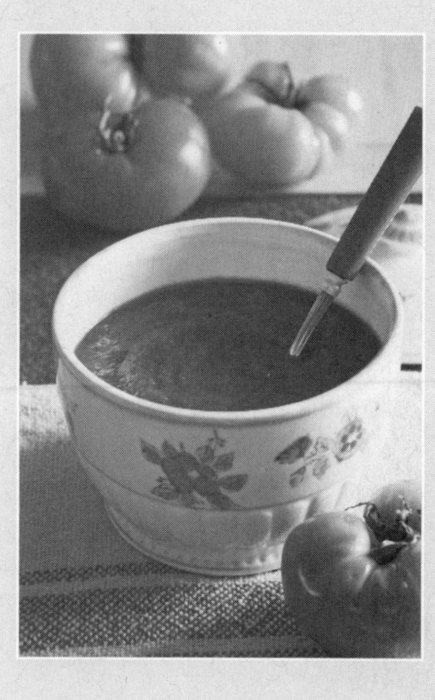

Fig. 10.26: Credit: Karen K. Will

Classic Pesto

The perfect accompaniment to tomatoes, summers wouldn't be complete without at least a half dozen pesto-paloozas in the kitchen. Start by growing a huge container or bed of basil in the summer. Harvest the leaves every week and keep that food processor whirring until you've got a freezer full of little gleaming green jars. You'll smile each time you pull one out to make pizza.

Makes 1 cup (½ pint)

⅓ cup toasted pine nuts
2 garlic cloves
2 cups packed basil leaves, washed and dried
⅓ cup extra virgin olive oil
¼ teaspoon salt
⅛ teaspoon black pepper
⅓ cup grated Parmesan cheese (optional)
½ teaspoon white vinegar or lemon juice (optional)

1. Combine the nuts and garlic in the bowl of a food processor. Pulse until crumbly.
2. Add the basil, oil, salt, pepper, and cheese. Process until smooth, scraping down the bowl once or twice.
3. Stir in the vinegar or lemon juice which will brighten the flavor and maintain the bright green color of the pesto while in storage.

Use within a week, or freeze in ½ pint plastic freezer canning jars (the Ball jars with the purple lids are perfect).

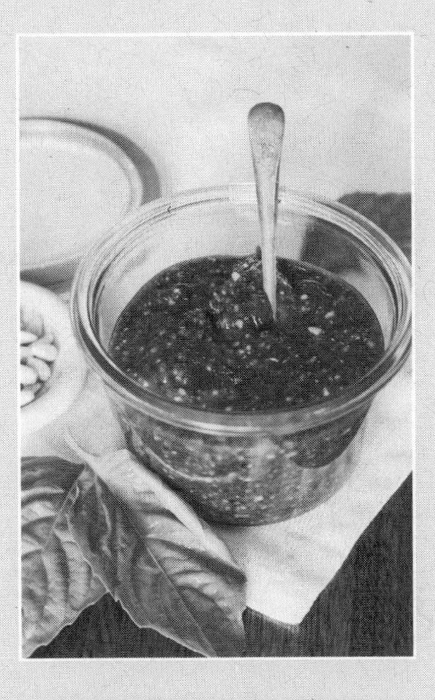

Fig. 10.27: Credit: Karen K. Will

Chapter 11

Organic, Handmade Kitchen Upgrade

We know that the kitchen is the "heart of the home," and never is this more true than with homesteaders. Every day, everything happens here ... from cooking and canning, preserving and fermenting, homework, cocktails, and heart-to-hearts with family and friends. The kitchen is the most active, yet comforting room of the house.

Home cooking fell out of favor sometime in the mid part of the 20th-century, partly because of the notion that kitchen work was "drudgery" and that a woman's time was better spent doing more important things. Labor-saving devices were invented to cut the time spent preparing a meal. Some, like the electric mixer and food processor, made us more productive; others, like the microwave oven, made us lazy. Now, in the 21st-century, our country is experiencing a welcome resurgence of home cooking. From kitchen gardens and farmers' markets to food magazines and cooking shows, it's evident that more and more people are in tune with what's in their food and are taking an active role in preparing it.

Since the kitchen is a room in which you'll spend so much time, it's not frivolous to make it a place where you're comfortable and where you can work efficiently — heck, even a place you will love to be! By making your kitchen the heart of the home with efficient workspaces, tools, appliances, and beautiful yet functional objects, you are celebrating the kitchen as a place of production and enjoyment. A simple, quality makeover will also make it possible for you to run a home-based food

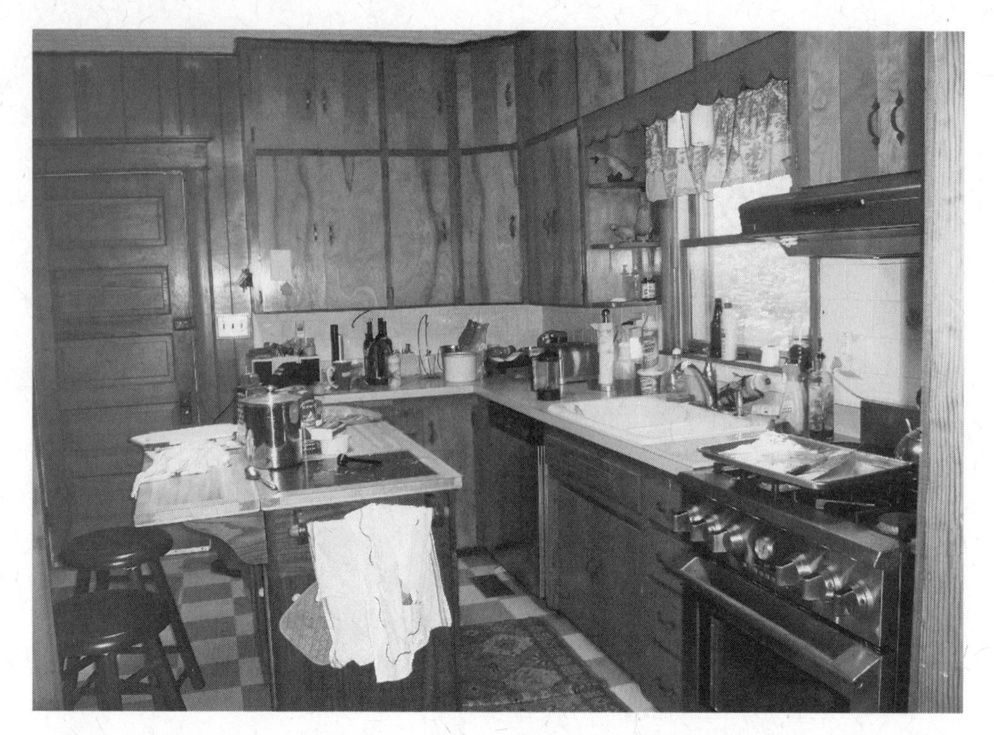

Fig. 11.1: *Begin your makeover by using your old kitchen — a lot. We lived with our outdated, dark kitchen for a couple of years and got to know exactly what we wanted to change.*
CREDIT: KAREN K. WILL

business there (discussed in Chapter 12).

Once you've been bitten by the farmhouse bug, features like an "upgraded" kitchen with granite countertops, stainless steel appliances, and custom cabinetry sometimes fly out the window in favor of a working septic system, a new roof, and functional outbuildings. Everyone's priorities will be different, especially when you move into a "fixer," but that doesn't mean you can't have an updated kitchen with modern conveniences to make your homesteading life easier and more productive.

It is possible to make over a farmhouse kitchen into a homesteader's delight without spending a fortune, as long as you're willing to put on your thinking cap and roll up your sleeves. Magic doesn't happen with hired contractors or designers; it happens in your mind and through your hands because your home is uniquely yours — not anyone else's — and frankly, no one else cares about it more than you. When it comes to renovations, don't think about how to "save money," think instead about how much you simply won't have to spend by proceeding in a frugal, yet intelligent manner.

Our philosophy when planning any project is: "As time and money allow." We commence projects, from fencing to kitchen makeovers, when we have enough money in the bank to pay for materials (and labor, if needed). To keep things in perspective, we very consciously do not use credit cards or home equity lines of credit. When we have to shell out cash money to complete a project, it keeps us grounded in reality so we don't drift into the fantasy world of credit. With all the kitchen options available these days, it would be no trick to spend $25,000 on custom cabinets. With a convenient home equity line of credit, those cabinets are more likely to sound like an affordable luxury; however, if you're writing a check from your checking account for cabinets, does it still seem reasonable — or even possible? Custom cabinets and granite countertops would be lovely additions if you have sufficient discretionary income, but if you don't, you've got to explore other options.

When we turned to rehabbing the kitchen in our 106-year-old, foursquare Kansas farmhouse, we didn't hire a designer or a contractor. We did what we always do ... we opened the DIY books. If you'd like to try an organic, affordable approach to making over your kitchen, here is a suggested one-year makeover plan based on our experience:

Step 1: Begin by *using* your kitchen — a lot. If you've just moved in, wait at least a year before demolishing the space, so you can figure out what works really well — and what doesn't. Think about where you can reach things easily — like all your baking trays and tins — and what items should be closest at hand. Is the divider in the double sink burdensome when you're trying to wash the big Crock Pot or a roasting pan? Do you never use that one corner of the kitchen because it's so poorly lit? Does the refrigerator stick out too far in the room, impeding traffic? Is the floor always dirty because it's light colored — and you have four dogs going in and out all day? Do you hate having to get out a stepladder to reach those upper cabinets? Think about the time and headaches you'd save by addressing these problems with a kitchen makeover. You want the kitchen to be the most functional, low maintenance, inviting, and productive room of the house, so start thinking that way.

Step 2: Next, pore over home magazines. The British periodicals like *Country Living, Country Homes & Interiors,* etc., seem to be much more practical and authentic than American "country" magazines with their New York version of "country style." Tear out pages showing kitchens you like, and keep a file. Look at colors, appliances, light fixtures, islands, hardware, unique features, fabrics, etc., and start to develop a realistic vision for your kitchen based on what already exists there. Come up with ideas for some minor things you could do to make it more functional and attractive. Buy or check out a book or two devoted to kitchen remodels. *500 Kitchen Ideas* by the editors of *Country Living* had quite a few dog-eared pages leading up to our makeover; it provided endless inspiration.

Step 3: Discuss with your spouse or partner, or just ask yourself, what you're willing to buy, build, or modify to achieve your vision for the room. If the kitchen needs new appliances, obviously that will be a big portion of the budget. If you have to spend most of your budget on appliances, well, that might just mean you'll have to work wonders with paint on everything else. Are you ambitious and handy? Perhaps you could craft a new kitchen island yourself (see Chapter 5) or install poured concrete countertops. Painting can be done by even the least-handy people with the right tools and supplies; count on doing that yourself.

Step 4: Research the options and prices for appliances, flooring, countertops, cabinets, sinks and faucets, hardware — and everything else. This knowledge will serve as a reality check — a baseline from which to plan and work. When you discover that granite countertops run $50 per square foot and up, you'll start to think in terms of "how can I achieve a similar look for much less money?" Could the answer be the aforementioned poured concrete countertops? You get the picture.

Step 5: Decide what the logical order of events should be. It doesn't make any sense to rip out an old vinyl floor straight away if you'll be painting and demolishing later; a new floor would surely get damaged in the process. Leave the old floor in place until the end, and you won't have to worry about spilling paint or gouging it with the refrigerator as you drag it off. You should also strategize about how to continue using your kitchen even while the major work — like cabinet and countertop installation — is happening. Rural living is no fun without a kitchen because there isn't a take-out restaurant or pizza delivery guy for miles and miles.

Step 6: When it's time to begin in earnest, schedule the work in phases. Don't think you're going to detonate the kitchen and put it all back together in a weekend, or even a week. Assume everything will take much longer than you think. Because you care about the outcome, you'll be paying attention to detail — which usually takes more time.

For example, set a goal for yourself to get one upper bank of cabinets painted over a long weekend. Saturday (or Friday night, if you're up for it) involves cleaning out the cabinets and finding a place to put all that stuff (tip: keep it accessible in boxes or flats so you can use the contents in the meantime, yet move everything around, if necessary). Next, you'll need to remove the cabinet doors and hardware, sand the surfaces, prime the doors and bases, and allow that to dry. Then, paint the doors and bases, allow that to dry, put all the hardware back on, and finally, put the stuff back in the cabinets, which of course, all needs to be sorted and arranged. You didn't think about all those steps, did you? Depending on the size of your kitchen and extent of your makeover, there could be five, six, seven or more phases to this work. And, don't forget about all the unexpected hurdles you may need to jump along the way. There could be wallpaper under that existing paint that will need to be dealt with in some way. There could be mold or mildew, wobbly doors, or drawers that stick. You could uncover evidence of mice (or worse!). Take the time to fix problems that crop up. There's literally no time like the present — when everything is ripped up and your tools and materials are out. Be patient, and stay the course!

Step 7: As you're patiently working through your phases, be on the lookout for unique pieces that could serve some purpose in your new kitchen. Trawl antique shops, or even the barn rafters, for items that can be salvaged or repurposed — even if for wall art. Items that have some history will give your kitchen a sense of place by connecting it to your land, your county, your state, your heart. Perhaps that old apple crate from the orchard down the road could serve as a bookshelf for your cookbooks. That rusty galvanized top to a cream separator that you found in the barn loft could be repurposed into a unique light fixture with a few inexpensive parts from the hardware store (see Pg. 220–221, *Salvaged Light Fixture*).

Fig. 11.2: *Make sure to plan for hurdles along the way ... wallpaper under paint, peeling plaster, and wobbly doors. Fix these problems now instead of doing a rush job.* Credit: Karen K. Will

Unfortunately with this plan, you won't get the big home-show "reveal" that's always so exciting on TV, but you'll get to witness your kitchen's evolution, in small (but sometimes large) increments, over time. The fun part of this plan is that your kitchen will probably never be completely done. You'll always find some antique knick or knack to add, or a door that could be painted a different color (inspired by a magazine picture you come upon), or a

Just Organize It

When devising the master plan for your kitchen remodel, don't overlook organization. Myriad systems exist for keeping things neat and ready at hand like pull-out baskets and shelves, lazy susans, drawer organizers, door storage, tiered spice racks, tray dividers for keeping upright things like baking sheets and muffin tins, pantry organizers and special units to maximize corner space, waste and recycling containers, sink front tip-out trays, and pot racks.

light fixture that could be swapped out. What we love about our kitchen is that it doesn't look like someone stepped in, waved a magic wand and poof ... the kitchen was "renovated." It was built with our own hands, piece by piece over a few years, and the end result is nice, clean, and comfortable, with vintage charm. Similarly, your kitchen can grow organically — just like everything else on your land — and you'll be proud that your head and hands were responsible for the changes.

Nuts and Bolts of the Kitchen

Life in a busy, productive farmhouse requires help in every conceivable form. In the case of a kitchen renovation, help comes in the form of low-maintenance finishes. You know in your heart that you just don't have the time to do labor-intensive, meticulous cleaning — or to waste time fretting about not getting it done. As you're considering your options, remember that the goal is to make mundane life (like cleaning) easier so you can concentrate on the myriad other projects that are infinitely more interesting and productive.

Appliances

Everyone has baggage, and when it comes to kitchen appliances, those bags are usually in various stages of wear. Fortunately, the refrigerator and stove don't usually conspire to bite the dust at the same time, so you may find yourself with appliances of assorted vintages. Assess what you have in good working order and what really needs to be replaced. You can sell unwanted or gently used appliances on Craigslist "as-is," and put the money toward the purchase of new items; or, you can donate them to a charitable organization (like Habitat for Humanity) for a tax deduction. An option for mismatched appliances is appliance paint. If you've got a white dishwasher, and a black stove and refrigerator, just spray paint the dishwasher black and presto! Now everything matches. But let's not get caught up in matchy-matchy-ness.

If you've decided to spend hard-earned money on things that really matter — tools — then you're thinking smart. Tools, in the form of efficient kitchen appliances, will make your life easier and more productive. This is not the area in which to skimp. Your kitchen appliances will be used to death in a busy kitchen with constant activity, so buy the best you can reasonably afford and figure out how much money you have left for everything else. By purchasing sturdy, quality items that are up to the job at hand, you won't need to replace them for many years, which means you'll also be contributing less to the landfill.

Refrigerator

Once you've determined where it will reside in your kitchen, take measurements of the available space. Don't delude yourself into thinking there really is enough space (or money) for that Sub-Zero side-by-side if there isn't. A refrigerator that is even half an inch too big can be a major headache — it can create a cascading effect such that now your other appliances won't fit. Or walls have to be moved. Or closet space reconfigured. So, make sure there really is enough room for the refrigerator model you're thinking about. Also think about traffic patterns, door clearance, and how the refrigerator relates to the other appliances. Would the refrigerator be better situated elsewhere, where it doesn't stick so far out into the room? Could cabinetry be built around it to make use of empty spaces above or on the sides?

When it comes down to picking a model, don't always go for the current trend, and yes, there are definitely trends when it comes to kitchen appliances. Having a double-sided refrigerator up top is very nice and handy. But what about that single-drawer freezer on the bottom that becomes more like a trash bin because everything gets just dumped in there? Chances are, some food will be overlooked and wasted. We know to expect an excavation project when we go to get something from our chest freezer, but for a kitchen freezer, we want to get in, find what we need easily, and get out in a snap.

Think about the food you'll store in the refrigerator and look for the model that will accommodate your lifestyle. Maybe you don't drink canned soda, so you don't need that built-in soda-can dispenser on the top shelf. If you're a devoted from-scratch cook, you probably won't need the (frozen) "pizza storage" slot in the freezer either. A better choice might be a model with huge vegetable storage bins (with humidity control) and pull-out shelves. Apply some critical thinking to the process.

Chest Freezer

This appliance is one of the most critical when it comes to rural living, especially if you raise meat animals, or buy them from your neighbors. A chest freezer opens up so many food-acquisition and storage possibilities. Whole or sides of animals are much cheaper to buy than individual cuts at the grocery store. A large chest freezer also allows you to make and store all kinds of produce and byproducts: such as berries and pesto during the summer; bags of pig fat from which to render lard; and large, bulky items like whole broiler chickens. A freezer full of grassfed beef, lamb, pork, and chicken is like money in the bank.

A chest freezer is highly efficient for two reasons: 1) Cold air sinks, so when the freezer is opened from above, cold air doesn't pour out as it does with an upright model. 2) Contents piled in a chest freezer create one thermal mass, rather than individual pieces around which cold air circulates; when the freezer is opened and closed, the mass hardly notices the intrusion, and it doesn't take much energy at all to replace the cold air.

Determine the ideal place for the freezer that's relatively close to the kitchen, yet is clean and rodent-proof. Ours is in the laundry room, just off the mudroom and kitchen. It's out of sight, yet still in the house and easily accessible. Measure the footprint of your intended space, keeping in mind the clearance needed for opening the top door. There aren't exhaustive options when it comes to chest freezers — some have sliding trays that make it a little easier to find what you're looking for; some lock (for security and as a deterrent to children trying to climb inside); but by nature, a chest freezer's ease of use is limited by its (efficient) design.

Optional: Egg Fridge

A fun addition to our laundry room, just off the mudroom and kitchen was a mini refrigerator we dubbed the "egg fridge." We had the problem of eggs piling up and taking over the kitchen refrigerator, especially during the summer, when 18 eggs per day is common. So, we spent about $100 on a mini fridge dedicated to storing the eggs we intended to sell. It also serves as a satellite fridge for things like bottles of beer and wine that we don't want taking up precious space in the main unit. A mini satellite fridge in the garage, barn, or mudroom can offer a little peace of mind to the harried homesteader.

Range

When you cook three squares a day, and perhaps even run a food-based business from your kitchen, you'll need your range to do more than sit and look pretty. An oven and stove will need to be able to run for long periods of time at high temperatures and produce consistent results time and again. Daily cooking will certainly put this appliance to the test. Be selective when it comes to purchasing your oven and stove — this is where the actual act of cooking happens; nothing is more critical to the finished product (other than the ingredients).

In our kitchen, we use a Kenmore Pro dual-fuel (electric and propane) range. It's similar to the professional-style Wolf or Viking ranges, yet it cost significantly less. This unit is the most important factor in keeping our home-based baking business in the black. The oven can operate at high temperatures (475°F for bread) for hours; the oven chamber is heavily insulated and is designed to retain and stabilize heat, so it produces consistent results — which is key because we make and sell baked goods every week. Even if you're not planning on running a food business from home, you might want to consider the possibilities. What business might you run — if only you had the right tools (a good stove, for instance)? Buy the best you can afford. Invest in equipment that will get the job done.

Cabinets

Kitchen remodeling goes hand-in-hand with watching scores of DIY shows on TV. Good, eye-rolling television is certainly made when some young house-hunting couple declares "this kitchen needs to be gutted!" Or, "I'm not crazy about these [dark, light, old, non-custom, what-have-you] cabinets." The sentiment reeks of primadonna-dom, and that's not what we're going for here (even if we could afford it). If cabinets are sturdy and well built, regardless of how "ugly" they are initially, a facelift is a relatively easy thing to do. Hardwood, custom-made cabinets are probably the single biggest budget item in a kitchen remodel, so start by removing the words "custom cabinetry" from your vocabulary and get your creative juices working overtime.

Cabinetry sets the tone for a kitchen. It usually dominates the walls, so color and finish is a big consideration. When we moved in, our farmhouse kitchen, which had been formally remodeled sometime in the 1970s, had sturdy, flat-paneled custom-built cabinets lining the walls all the way up to the 9-foot ceiling height. Constructed of fir plywood, they were finished with a simple coating of glossy varnish. This quintessential 1970s style was in vogue then, but not so much today. The walls were wood-paneled in the same mid tone, making the room extremely dark; we found it depressing. But the cabinets were custom built for that space (albeit in the 1970s); replacing them with custom cabinets that we actually liked would be extremely expensive. We decided that the best and most affordable option was to work with what was there.

The first step was to dress up the flat cabinet faces. Using a router, we created a groove down the middle of each door, creating a country-style "barn door" effect. Next, we lightly sanded, then primed and painted the cabinet bases and doors with bright white paint. The effect on the room was instantaneous — lightness was possible! On the route to white-cabinet nirvana, we found ancient, crumbling wallpaper inside the upper cabinets, unfinished

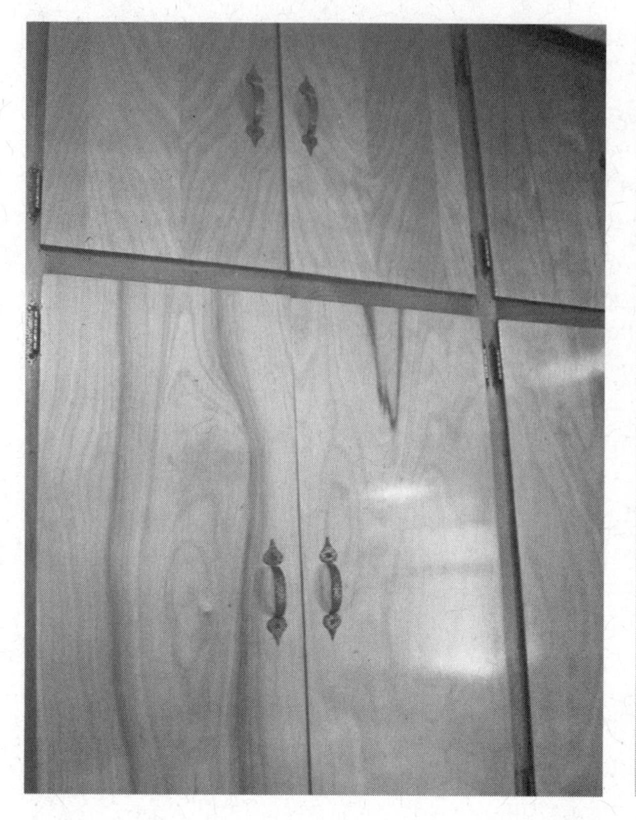

Fig. 11.3: *If cabinets are sturdy and well built, regardless of how "ugly" they are initially, a facelift is a relatively easy thing to do.* Credit: Karen K. Will

Fig. 11.4: *The finished cabinets were updated by routing a groove down the middle of each door, creating a country-style "barn door" effect.* Credit: Karen K. Will

plaster walls, and countless layers of shelf paper that presented a timeline of kitschy styles through the decades. The extra prep work foisted upon us involved peeling wallpaper and shelf liner, replastering sections of wall, and seal-foaming the tiniest cracks and crevices to keep out mice. It was messy work, but we threw ourselves into the process, convinced it was worth it, and laughed about how much it would have cost to hire someone to do it — if there was such a person.

Other possibilities for making over old cabinets are refacing (this is somewhat pricey when hired out, but doable yourself with a table saw and router), distressing, stenciling, faux finishing, glazing, or simply dressing them up with new knobs and pulls. If you decide to paint, take a bit of time to decide on color. Home improvement and paint stores sell little tester jars of paint, so you can do a color trial before committing to it. To test out various finishes — like distressing or stenciling — get a half sheet of plywood or even a 2 x 4 in roughly the same wood and finish style of your cabinets and go through the whole process of applying the finish you've decided on. This practice run will help you make sure you like the finish. Plus, it will give you some practice for doing the actual cabinets. Work out all the kinks *before* you tear up the kitchen. You don't want to spend lots of money on paint or special finishing materials, and then determine you don't like the results.

There are many tasks to get done before you get to the actual painting of the cabinets. This somewhat harsh reality shouldn't be taken lightly. As excruciating as prep work is, always take the time to do it thoroughly — it will contribute to a long-lasting finish and prevent you from having to repeat your work.

For sure, you'll need to do the following: Remove everything from the cabinets; remove doors and hardware; remove old wallpaper, shelf liner, peeling paint, and anything else that would get in the way. Seal cracks and crevices with expandable foam or caulk; clean all surfaces with vinegar and water; lightly sand the cabinet bases and doors to rough up the surface and give the paint something to cling to; wipe cabinets clean with tack cloth to remove sawdust.

When it's time to begin painting, set up an assembly line with saw horses or tables where you can line up all the cabinet doors, paint them, then flip them over and do the other side. If you're sure that rain isn't in the forecast, you can do this outside (as long as it's not humid or there aren't a million gnats flying around), but the garage or

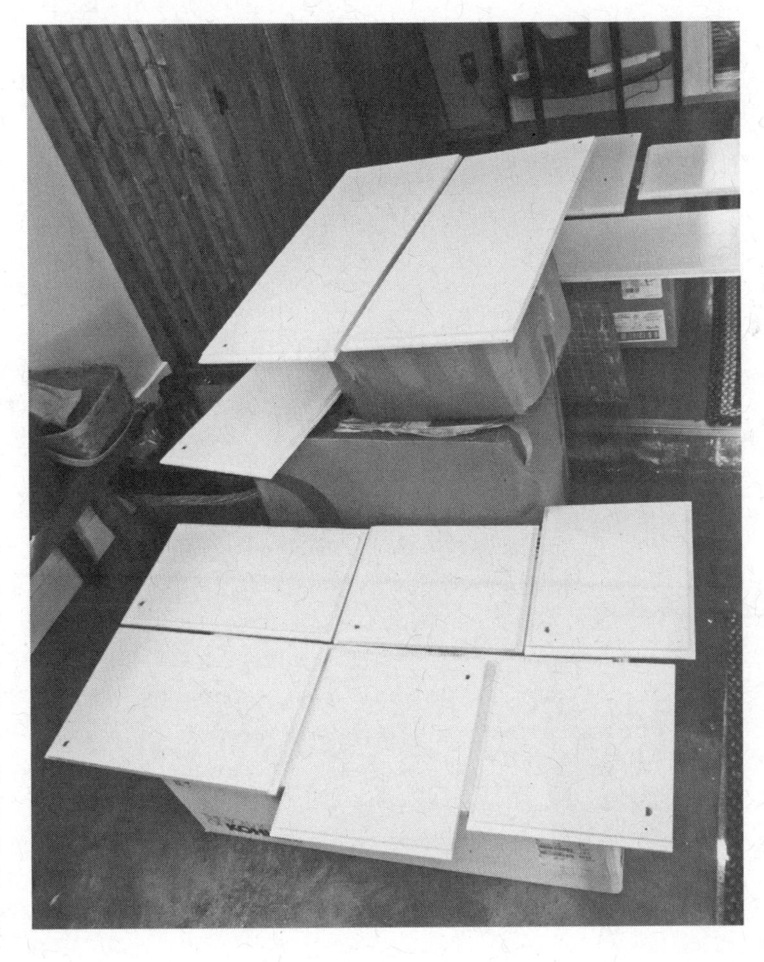

Fig. 11.5: *When it's time to begin painting, set up an assembly line with saw horses, tables (or whatever you can find), where you can line up all the cabinet doors, paint them, then flip them over and do the other side.* CREDIT: KAREN K. WILL

barn would be a better choice. Start by priming cabinet bases and doors with a high-quality primer. Somewhat new to the market are "paint and primer in one" formulations. They offer instant gratification because you get to see the color right away. But, even though the can says you won't need to, you'll probably need to apply two coats. So, "paint and primer in one" can end up being the more expensive choice because primer is cheaper than paint (about half the price). One coat of good primer, plus one coat of good paint costs less than two coats of "paint and primer in one."

Purchase paint that's been specifically formulated for kitchens and bath — one that's durable and washable. If you want your cabinets to scream, choose a semi-gloss or high-gloss finish; if you want them to be more subdued, go with a satin finish. If you use good paint and a high-quality brush, you might actually find painting to be enjoyable; with the right tools, the job will flow with ease and look professional. Allow plenty of time for the paint to cure before reinstalling and restocking the cabinets. If you're not patient with this step, you're likely to put some nicks and gashes in the new finish — and they'll gnaw at you every time you see them. Remember, cabinets set the tone for the kitchen, so due diligence is required for this step.

Close the Door on Open Shelves

If you look through home magazines, you'll see that using "open shelving" as a replacement for enclosed cabinets is a popular style right now. It's a beautiful look that imparts an airiness to a kitchen that's delightful. If you're contemplating this arrangement, here's what I say: Stop right now, and think about what you're doing! Unless you plan to devote half your kitchen storage space to your charming collection of antique ironstone bowls and pitchers, or the only food you consume is brightly colored cans of San Marzano tomatoes and bottles of San Pellegrino, your kitchen will not look like the magazine kitchens. Think about what's in your kitchen cabinets right now. Now imagine ripping off all the doors. Do you really want to look at that pile of Tupperware or those stacks of mismatched coffee mugs every time you go into the kitchen? Is it feasible to factor in the package design of everything you buy at the grocery store so that it looks neat and tidy when displayed on open shelves? On top of the aesthetic detractions, consider how dirty items will get when exposed to the kitchen environment. Unless you use every piece every other day, items on open shelves are going to collect dust, and you'll have to wash everything before using it. No fun. Practical heads must prevail here — forget about open shelving!

Countertops

Countertops can be considered another tone-setting feature of the kitchen. For some people, it's all about the countertops, for others, it's all about the floors or appliances. Regardless of what you have on your countertops now, if you don't like them, there is a way you can refurbish them for very little money. It could be as easy as covering those countertops with linoleum tiles, or painting them with a "Countertop Transformations" kit, a multi-step process that can be completed in a weekend for about $250. Other options, ranging in price, include:

Wood

Wood is the epitome of a country farmhouse kitchen. It's abundant and natural; it's waterproof when sealed, easy to repair, and nicks and scratches only add character and appeal over the years. You can build and install wood countertops yourself with minimal DIY acumen. If purchased and installed by professionals, the cost is still less than for other natural materials like granite or marble.

Concrete

Concrete is a common and inexpensive material, and one that fits right in with the gritty farmer aesthetic. Concrete countertops are highly functional. Blemishes add to their character — you won't be fretting about ruining your countertops in any way when they're concrete. You can either hire this work done (it can take up to 10 days for a professional to do the installation) or you can do it yourself. The quick and easy method is called "cast in place," which involves simply building a form on top of your cabinets and pouring the concrete. Another, more complex method involves mold making, pouring, extensive polishing, and careful installation; it's best suited for fairly skilled DIYers. Books and videos are available on making your own concrete countertops; Quikrete makes a special "Countertop Mix" formulated for concrete countertop applications.

Stainless Steel

Though not the cheapest, stainless steel is a great option for many reasons: It's one of the sturdiest materials available, so you'll have a tough time damaging it, and it's non-porous, so it won't stain or absorb bacteria — a great bonus if you're considering a home-based food business. Stainless steel is heat resistant, so you can set hot pots and dishes directly onto the countertop without first having to search for a trivet — so it gives the functionality of a professional kitchen. Aesthetically speaking, stainless steel reflects light, so if your kitchen is small and/or dark, it can brighten up the space and make it more inviting.

Super-Easy Concrete Countertops

My innovative and frugal cousin, Laura Hathaway, figured out the easiest-yet method of creating a concrete countertop in her kitchen. While searching online, she found a product called Ardex CD (stands for Concrete Dressing; www.concretedressing.com), a Portland cement-based product specially formulated to resurface and repair concrete surfaces such as driveways, sidewalks, and patios. Used mostly in commercial applications, it's billed as a low-cost, revolutionary alternative to concrete replacement that provides "a durable new wear surface for foot and rubber tire traffic both outdoors and indoors." The product's claim to fame is that it will bond to anything — even glass — very quickly, so Laura figured she could easily cover her ceramic tile countertop, creating a new surface that better suited her style.

Ardex mixes with water and has the consistency of peanut butter. According to Laura, it has a working time of only about 20 minutes. While her husband handled the mixing of the product, Laura quickly troweled on the concrete (directly over the tile, and without using molds) in sections, always keeping a wet edge. After roughly covering the countertop, she ran a wet sponge over the cement to slightly smooth and level it. She even covered the vertical edges of the counter, giving the appearance of a thick concrete slab. After curing for just six hours (the manufacturer claims it can be walked on in two hours; driven over in six, and sealed the same day), she dry brushed on burnt umber paint and several coats of Varathane, lightly hand sanding in between coats.

She says next time she'd seal the concrete counters with epoxy instead of polyurethane (the Varathane), simply because it's more durable; five years later, there are some scratches and nicks in the finish.

The result is a rustic, somewhat industrial-looking countertop, done for about $250.

Fig. 11.6: Credit: Laura Hathaway

Laminate

For all the bad rap that laminate counters get, this material has really outgrown its ugly, awkward phase. Since there are so many colors and patterns available from companies like Wilsonart and Formica, and the cost is relatively low, laminate is a great choice when you're on a refurbishing-your-farmhouse-budget. If laminate is the most appealing option budget-wise, go for a solid color that will be versatile and stand up to the test of time — matte black laminate counters can take on the lovely look of soapstone. There's nothing worse than a busy pattern that seems dated just a few years later ... gingham shelf paper comes to mind. You have to be careful with laminate, though. You can't set hot pots on it, and you need to be somewhat gentle because laminate will chip and it shows scratches more easily than stone or tile — and repairs aren't easy.

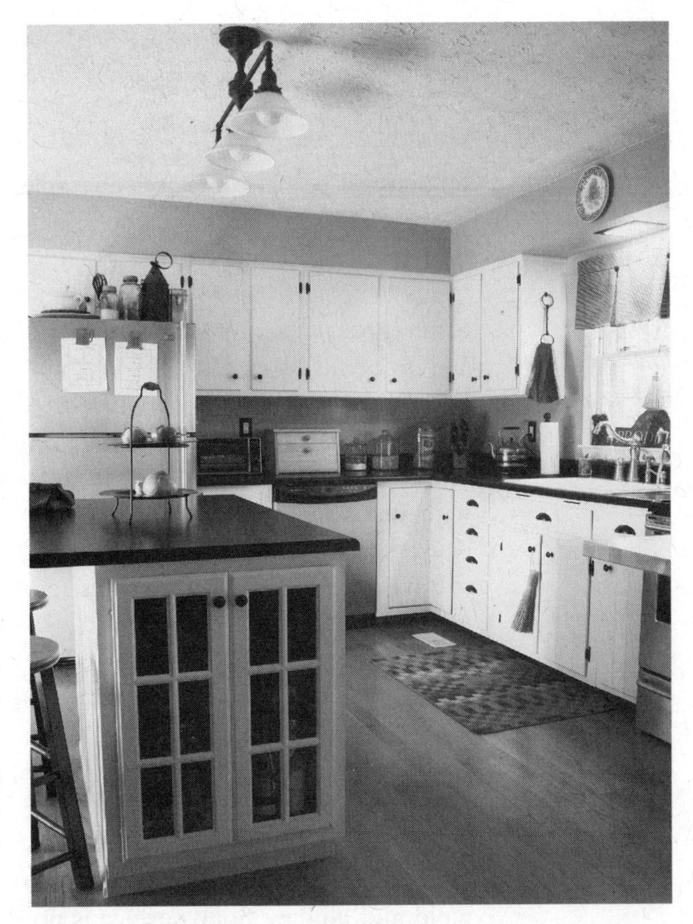

Fig. 11.7: *If laminate is the most appealing option budget-wise, go for a solid color that will be versatile and stand up to the test of time — matte black laminate counters can take on the lovely look of soapstone.* Credit: Karen K. Will

Granite

It's true that natural stone like granite is both expensive and nonrenewable, but you can't beat granite for its durability, resistance to scratches, stains, heat, mold, and mildew. And let's not forget its beauty. These are the reasons granite has set the bar for kitchen countertop material. Granite is not a DIY project; if you choose it, seek out an independent outfit that allows you to visit a slab yard and select a unique piece. Don't rely on the stock selections at home improvement stores (especially because prices are not much different). We found our countertop installer on Craigslist, and we had a perfect experience. However, it's not a bad idea to check Better Business

Bureau ratings and/or referrals before making a commitment.

Quartz

Engineered quartz countertops like Silestone and Zodiaq are made of natural quartz combined with polymer resins in order to give the stone a uniform appearance and strong, durable, and hygienic properties. These countertops are virtually maintenance-free. They are resistant to stains, scratches, and scorches, and they're easy to clean with just soap and water. Prices are equivalent to natural stone. You'll need to have quartz countertops professionally installed.

Solid Surface

Products like Corian are made from acrylic polymer (plastic) and other inorganic compounds. It's a popular material for countertops because it's seamless, easy to repair, and resistant to mold and mildew. Solid surface is not the least-expensive option, but it costs less than natural materials like stone (granite, marble, and quartz). Solid surface holds up to stains, scorches, and scratches without having its finish destroyed; blemishes can be repaired with a little sandpaper or a scouring pad and some cleanser.

Tile

Regardless of composition, tile is best suited for a backsplash rather than a countertop. All those grout lines will pick up bacteria and stains, and the uneven surface will annoy any cook over time. Glazed tile is durable, easy to clean, low maintenance, and impervious to stains and moisture; pair it with gray or beige grout (anything other than dirt-prone white), and you've got an extremely economical and reliable backsplash that you can install yourself in a weekend. Our backsplash is made from 5" x 3" glazed ceramic white "subway" tile that we purchased in-stock at a home improvement store.

You can make over an existing tile backsplash by painting it with oil-based paint. For a small but significant change, try painting your existing grout with grout colorant. It's very easy to apply and can give tile a whole different look and feel in a matter of minutes. Home improvement stores stock a few basic colors, but scores of colors can be found online from manufacturers like Mapei. Another route to go would be leaving it alone, and creating a color scheme for the rest of the kitchen around it.

Eco-friendly Materials

If your primary concern is using sustainable and renewable materials, consider countertop options like PaperStone, made from recycled paper and petroleum-free resins; recycled aluminum, fabricated from aluminum scrap and resins; or recycled glass tile.

Sink

There seems to be a love affair going on between every person who's remodeling a kitchen — from New York City to Topeka, Kansas — and the ubiquitous big "farmhouse sink." That single-basin, apron-front, genuine cast iron gem is ideal, for sure, with its ability to let you wash large, unwieldy roasting pans one day and a dirty dog the next. However, a farmhouse sink doesn't come cheap. And what, exactly, is the use of an apron front? Sure, it's charming, but it's certainly not necessary for a single-basin sink. Much more economical models exist in stainless steel, acrylic, or cast iron (sans the apron front). Search the Internet for "discount sink" or "discount plumbing fixtures" and you'll turn up dozens of options.

All sorts of high-quality faucets are also available online; look for all-metal cast construction, which will hold up longer than acrylic versions, and a high-arch spout — perfect for filling large pots.

Flooring

The name of the game when it comes to flooring and rugs for the kitchen is "low maintenance." We had a sage green and white vinyl checker-patterned floor in our kitchen that had average wear — stains and nicks — after five years. But ... because it was light colored, it showed every spill, every speck of dirt, and every muddy pawprint, of which there were plenty. It became a constant source of stress and frustration trying to maintain a modicum of cleanliness. A quick fix was to put down rugs and runners made from miraculous polypropylene, otherwise known as "indoor/outdoor" rugs, to absorb some of the detritus (for details, see the sidebar, "Falling in Love ... with a Rug").

In the past, homeowners pretty much had two choices: tile or linoleum. Today, the options have expanded considerably.

Wood

Though hardwood floors are the epitome of country style, they generally aren't considered practical for kitchens (due to the kitchen being a "wet zone"). However, wood flooring is popular for use in open floor plans (because it can seamlessly blend together the kitchen and family room) or when style calls for the warm and inviting nature of wood. If you're set on wood flooring in the kitchen, opt for factory pretreated wood with a highly durable sealer like polyurethane. When this type of flooring is manufactured, somewhere between six and ten coats of sealer are applied, making the finished product practically impenetrable — and so not subject to the expansion and contraction that happens when water meets wood. A properly sealed and installed wood floor is durable, easy to clean with a broom and "Swiffer" or "Bona" type mop, and they handily camouflage spills, dust bunnies, and tracked-in dirt — a godsend in a busy farmhouse.

If you decide to use a hardwood floor that was underneath some other

Falling in Love . . . with a Rug

In the spirit of working with existing materials, we could not replace our kitchen floor (yet another shade of white in a Kansas farmhouse!) in the early stages of our makeover. But every time our two terriers stampeded into the house from outside, they tracked dirt or mud on the white kitchen floor. I found myself either: a) stressing out and feeling like a bad housekeeper for having always-dirty floors; or b) constantly getting out the vacuum and mop to clean up the detritus. Neither was healthy or an effective use of time.

Fig. 11.8: Credit: Karen K. Will

The simple solution, after pondering for some time, was what retailers market as "indoor/outdoor" rugs. Whoever invented this magical material — polypropylene (a form of plastic sometimes made from recycled materials) — was a genius. And I'd bet it was a "dog person." The rugs are indestructible, nothing sticks to or stains them, and when they do get dirty (like when a dog has an accident), they can be taken outside, hosed off, and left to dry. In the spring or summer, drying takes about an hour, draped over a fence. The rugs never lose their shape or color, and they'll last for years.

The best part? Since manufacturers market these rugs for use outside, perhaps just for a single season, they're much cheaper than rugs sold for interior-only use. All this equals the "perfect storm" for a frugal farmhouse!

My favorite sources for polypropylene indoor/outdoor rugs are the online JCPenney catalog (I love their country-style braided runners); Overstock.com (hundreds of choices at discount prices); and Pottery Barn (their Kilim-style rugs are made from recycled soda bottles).

material, make sure to sand and finish the floor with multiple coats of polyurethane to protect the wood, especially in that "wet zone" known as the kitchen.

Installed hardwood flooring costs in the neighborhood of $4 to $10 per square foot, depending on the type of wood you select. Deals can be had on recycled wood from salvage yards and teardowns of older homes — a great way to save something from the landfill while giving your home a link to the past.

Concrete

Just as concrete is used for countertops these days, it can also be transformed into something beautiful for the floor. If your kitchen already has a concrete subfloor, existing flooring (like vinyl or tile) can be pulled up and the concrete refinished into a stand-alone floor. If you don't have a concrete subfloor, you can make a new concrete floor by installing thin slabs on top of the subfloor.

Concrete can be colored via an acid-staining process, called "etching," in which a mixture of hydrochloric acid reacts with the concrete. After staining, sealer and wax are applied, resulting in a lovely burnished sheen. Acid-stained concrete can mimic tile, marble, slate, or even hardwood, depending on how the stain is applied. Etching is a relatively simple DIY project, and supplies are available at home improvement stores.

Sealed concrete kitchen floors have advantages — they're non-porous, so

No Stone Unturned

Any time you can reuse or repurpose something old and give it new, meaningful life in your kitchen, thus saving it from the landfill, you're truly being "green." With that and a low budget in mind, your best bets for kitchen goodies are rural antique stores; rural because prices are definitely lower than big-city antique stores with their high overhead. Rural antique stores will also have more of the "real deal" finds like milk cans and funnels that were trawled from farm auctions and estate sales from the area. Other sources include flea markets, thrift stores, yard sales, farm auctions, and estate sales. Online sources like Craigslist (check daily under the appliances, antiques, furniture, and household categories for items to reuse and repurpose, like a kitchen island); eBay; and Etsy are all mercantiles of unique kitchen items. Overstock.com is a mainstream, yet great discount source for rugs and most all home furnishings; discount hardware and plumbing sites like KnobDepot.com (great for cabinet hardware), HardwareHut.com (lots of hard-to-find niche hardware and switchplates), and FaucetDirect.com (kitchen sinks and faucets), are invaluable when taking on a kitchen remodel yourself.

they won't absorb spills, bacteria, or allergens. Concrete is also cool ... meaning it doesn't heat up very quickly, so the kitchen can be kept at a comfortable temperature — great when running a baking marathon. However, think about how hard it is to stand on concrete for long periods of time (ever been to a trade show?). If you'll be standing in the kitchen for hours at a time, concrete will take a toll on your feet, legs, and back. Anti-fatigue mats or rugs will help, but durability and low maintenance have a trade-off, in this case.

The costs for installing and staining a concrete floor are reasonable, about $3 to $10 per square foot, depending on what type of stain and sealant are used.

Linoleum

Perceptions about linoleum run the gamut from old-fashioned to drab, but you might be surprised to hear that genuine linoleum is actually a "green," environmentally friendly flooring made from all-natural ingredients — a mix of linseed oil, cork dust, wood flour, tree resins, ground limestone, and pigments — the same recipe concocted and patented by Englishman Frederick Walton in 1863. Linoleum is not related to vinyl in any way, and it technically surpasses vinyl because, whereas most vinyl patterns are printed on the surface, linoleum's colors bleed all the way through, gently wearing over time and gradually revealing new levels of color.

Linoleum has natural bactericidal qualities, and it's quite durable; floors can last 30 or even 40 years. But, because it's porous, it needs diligent maintenance — new floors should be sealed with one or two coats of acrylic, then recoated once a year after that to keep them looking fresh and protected from stains.

Made in sheets or tiles and stuck to the floor with adhesive, linoleum tiles can be a DIY project (if directions are followed carefully!). The cost is easy too: from $3.75 per square foot for tiles to $4.75 per square foot for sheeting (professionally installed). It's water-resistant, comfortable underfoot, and it's available in hundreds of vibrant colors. Linoleum is also easily repairable with wood glue and scrapings from an extra piece — perfect for when you drop a knife.

Vinyl

Like laminate countertops, vinyl flooring has come a long way. Patterns exist that mimic hardwood, stone, tile, and other natural materials. Though vinyl is ridiculously inexpensive — starting at less than $1 per square foot — it's not perfect. Vinyl is a synthetic product made of chlorinated petrochemicals, and it will knick, stain, and burn relatively easily. It also can't be repaired in the way linoleum can.

However, self-adhesive floor tiles are an easy DIY project — a quick, cheap makeover that can be completed in a day or two.

Tile

Thanks to its strength and resiliency to staining and wear, ceramic, porcelain, slate, travertine, terra cotta, or saltillo earthenware tile are classic choices for wet zones and areas that get a lot of foot traffic. Like wood, the beautiful, natural variation in tile gives a rustic, lived-in and comfortable feel to a room — in fact, tile is now being made that looks identical to wood! While not extremely impact resistant, the wearability of tile is the reason for its durability. When breakage happens, repairing tile is about as easy as fixing other types of flooring. Remove the broken pieces, clean the area, reinstall replacement tiles and regrout the repaired area. Don't neglect to keep extra tiles on hand for such repairs.

Installing earthenware tile is easy for the average do-it-yourselfer. In fact, home improvement chains regularly offer free classes on how to do it. Basically, it involves prepping the area, cutting and installing the tile, and grouting. It's a project that can be completely finished in a weekend.

Regardless of type, tile usually comes in a variety of sizes, from square 4″ x 4″, 6″ x 6″, and up to 24″ x 24″. Cost is another reason why tile flooring is so popular — ceramic can be had for less than $1 per square foot.

Brick

If tile is the most logical choice for your kitchen, but you want something a little more unique, consider brick, with its cozy, vintage farmhouse feel. Brick flooring pavers are thin tiles of real clay that look like aged brick but are installed and maintained just like tile (and cost about the same too). They come in lots of colors and textures and can be laid in any number of patterns. Most importantly, they're practically indestructible in a busy kitchen and don't show any signs of dirt.

Hardware

Cabinet hardware is the easiest thing in the world to change. Just by swapping out those gold-toned or otherwise-dated knobs and pulls with something more to your taste, you'll instantly change the whole look of the cabinets, and thus, the kitchen. If the existing hardware is beyond rehabbing, go online to shop for new pieces. Just by doing a Google search for "discount cabinet hardware" or something similar, you'll turn up multiple sites that offer everything you'd find in the stores, often with many more color and finish options, and sometimes for much lower prices. When you have 72 cabinet knobs that need replacing, you'll appreciate that $1 to $2 lower price!

Before buying, make sure to measure the existing hardware (circumference of the knob, distance it protrudes from the cabinet face) and compare it to the measurements of the style you like. Don't skip this step. Pictures on a computer screen can be deceiving. You don't want any disappointing surprises when they arrive in the mail. If you're very cautious or frugal, you could order just one or two knobs at first. Install them and live with them for a few days to see if you really like them before ordering the whole lot.

If you want to rehab the old knobs, pulls, or even hinges (which are notoriously hard to replace on older houses because the back spacing on modern hinges is now standardized), it can be a really fun project involving paint. Metallic finishes in antiqued bronze, wrought iron, hammered copper, pewter, or silver all can transform hardware beautifully. Head down to the hardware store and survey the options. Make sure to choose a high-quality brand suitable for use on metal (or whatever the material you're painting) since you'll need the finish to hold up to constant handling. Don't forget the primer either.

First, remove knobs from the doors and drawers and lightly scratch up the surface of each knob with a piece of sandpaper, then rub clean with tack cloth. To make knob painting easy — and the finish flawless — make a painting station first: Get a quarter sheet of plywood or an old scrap of wood that is at least ½-inch thick. Pound finishing nails three inches apart all over the wood — enough for all your knobs

Salvaged Light Fixture

One of the final steps in our ongoing kitchen renovation was to fill the (electrical) box left by a ceiling fan, which we had moved because its blades were hitting a cabinet door when it was open. At an antique store, we stumbled upon an old milk can funnel that had a charming rust veneer over galvanized steel. It struck us as a perfect lampshade, so, after parting with $6 for it, the project of turning it into a kitchen light fixture began to take shape.

The vintage milk can funnel is made from tin-plated mild-steel, which is easy to modify as needed. I chose a gray powder-coated, adjustable lamp head for the bulb base. This base has a ½-inch NPT thread at the end opposite the bulb socket. I used the supplied locknut to support a large galvanized fender washer I found in my parts box. The washer keeps the funnel in place and helps keep its soft perforated steel from deforming under the funnel's weight.

I threaded the end of the lamp head directly into a ½-inch galvanized pipe coupling. I used tin snips to open the end of the funnel enough to make ☞

room. My snipping was sufficiently close that I didn't need a fender washer directly beneath the coupling. Tightening up the coupling holds the funnel securely in place. To the other end of the coupling, I attached a 12-inch-long piece of ½-inch galvanized pipe, threaded on both ends (a 12-inch nipple). The end opposite the funnel threads into the lamp base that goes with the lamp head I used inside the funnel. That base is screwed to the ceiling box.

Since we wanted this light to be operable separately from the wall-switched main kitchen light, as was the case with the old fan, I left that side circuit wired hot and installed a pull-chain switch inside the lamp base.

Since the lamp base is quite angular and modern looking, we decided to cover it with a canopy. We couldn't find any canopies at the local electrical supply place that suited perfectly, so we just bought an inexpensive one and modified it. First, I used a large twist bit in my drill and did some filing to open up the hole in the center to accommodate the ½-inch galvanized pipe. Second, I drilled and filed out a hole to allow the pull switch to see the light of day. I noticed that the canopy required a couple of screws to be threaded into the fixture to which it was designed to be attached. To work around this problem, I drilled and tapped the lamp mounting base for screws that would fasten the canopy in place. Those screws are beneath the two acorn nuts.

As with any electrical work, don't take on more than you are knowledgeable enough to handle. Simple wiring and lamp installation is fairly easy if you understand the fundamentals and have excellent resources at hand. Study the appropriate sections in home improvement books and online to ensure that your work is up to code and most importantly, safe. Our homemade milk can funnel light serves its purpose wonderfully and its rustic nature fits the kitchen perfectly. If you prefer a more uniform look, you could paint all the components in the same color; be sure to let the paint cure thoroughly before assembling.

Fig. 11.9: CREDIT: KAREN K. WILL

(or work in stages). Through its screw hole, perch a knob on each nail.

Next, spray primer on the knobs by holding the can six to seven inches away and spraying evenly. (The three-inch spacing between nails will allow you to get all the way around each knob and thoroughly coat with paint.) Make sure each knob is opaque with primer, but don't overdo it. Follow the instructions on your primer can, and paint with the finish color when the window of time is right — don't rush it, but don't wait too long either.

When it's time to topcoat, spray each knob lightly, at the recommended distance, in several coats until done. This is the important part: Let the paint cure completely before replacing the knobs. Because the knobs will be handled constantly, a soft, uncured finish will instantly scratch or gouge. Leave the knobs on their little rack and turn your attention to another project until a few days have passed.

Island

If the kitchen is the heart of the home, then the kitchen island is the heartbeat. A kitchen revolves around the island, and it serves the important function of expanding counter space. The kitchen island is the perfect candidate for reusing or repurposing an old or antique

Fig. 11.10: *Our kitchen after "phase one" was complete. A handcrafted kitchen island, a "tinned" (painted) ceiling, and hardwood floors were still on the horizon.*
CREDIT: KAREN K. WILL

furniture piece, given it's sturdy and up to the task.

First, decide whether you want the island to be movable or fixed. Should it be functional as a counter with stools, or simply as a food-prep area? Before you hit the flea markets and antique shops, take measurements of the space available. Figure out what your ideal measurements are for the island, so you know how much wiggle room you have. Consider all kinds of pieces like old bars or shop counters, library tables, long harvest tables, stainless steel carts, vintage sideboards or credenzas — anything with the right dimensions. If it's not the proper height for chopping duty, you can add locking casters or wooden tapered or ball feet (available at home improvement stores) to raise the height by a few inches. If the top isn't quite right, alter it with a new or reused piece of marble (ideal for baking), butcher block, concrete, or the old stand-by — paint.

Fig. 11.11: *Here's that "before" photo again. What a difference!* CREDIT: KAREN K. WILL

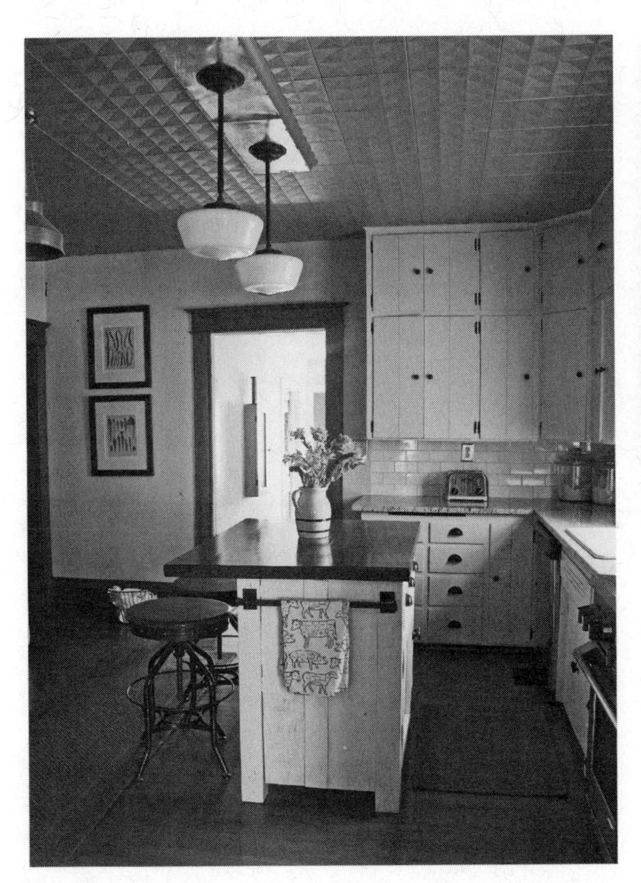

Fig. 11.12: *Finally, our farmhouse kitchen seems finished.* CREDIT: KAREN K. WILL

Chapter 12

Home-based Food Business

The homesteading life is centered around the home ... producing things at home, finding entertainment at home, living off the land. With life centered on the home and farm, we're reluctant to be employed away from it. And, because we were already engaged in production, it made perfect sense to expand our skills into some kind of home-based business. It might be for you, too. It wasn't that long ago that farm women always earned a little side cash from their home-based endeavors (remember "egg money"?). A few hours a day — or week — can turn into the week's grocery money or pay some of the monthly utility bills. Working from home also allows you the luxury of not completely detaching from other home-based duties like homeschooling, tending to livestock or pets, cleaning, cooking, and laundry. It's possible to make some cash doing something you love from home.

If this idea appeals to you, take inventory of your talents, interests, and limitations to determine what you could turn into a business. Also ask yourself how much money you want to put (or can put) into starting a small business, as well as how much you'd like to make (be reasonable with this one!). Look around — what are you doing already that others might pay for? If your small flock of chickens lays good eggs, could you expand the flock and sell the extras? If you have a green thumb and produce bushels of garden vegetables every year, could you plant even more and start up a roadside stand, a booth at the farmers' market, or even a small CSA? If you

The Local Loaf

When I decided to get off the office-job merry-go-round and get serious about working from home, I took stock of my talents and limitations. Besides writing and photography, I loved to cook and had recently discovered the art of bread making. One day, as my husband chewed a piece of fresh-baked bread, he remarked "This is so good, you could sell it ... Seriously." So, we began talking about how my baking from home could work as a business endeavor. I knew baking for profit was a tried-and-true business model for many homemakers before me, so I proudly took the idea to heart.

Since Hank works for a mid-size publishing firm in Topeka that specializes in homesteading and rural-living magazines, I decided to test market my bread with that crowd, being the perfect target market. So, I baked four loaves of bread early one morning and sent them to work with Hank, along with cutting boards, good bread knives, and butter, of course. He asked tasters to email me with feedback, and I took the opportunity to ask them if they'd pay for such a product. The response was overwhelmingly positive, so I quickly put together a business plan in my head, and decided to strike while the iron was hot.

After determining that selling baked goods from my farm kitchen was perfectly legal in the state of Kansas (being a "low risk" food), I collected the names and email addresses of everyone who stated they were interested, and sent a group email offering four or five different breads, along with prices and protocol — the call for orders goes out on Wednesday via email; customers respond with orders on Wednesday or Thursday (up until ☞

Fig. 12.1:
CREDIT: KAREN K. WILL

noon, which I established as the "cut off time"); bread is baked on Friday morning and delivered the same day by my husband at work. Cash or checks were accepted.

I figured on baking four loaves of bread in a morning — I had two cast iron pots, and my oven could bake two at once, and there would be time enough in a morning for two batches. When I quickly received orders for six loaves the first week, I had to rethink my plan. I decided I wasn't going to turn away any orders, so I would just have to get up earlier and somehow manage. This was a business proposition after all, not just my hobby anymore.

I decided on a business name — The Local Loaf — and designed a label for my bread tags. State regulations require me to affix my name and complete contact information, in addition to a full list of ingredients to the bread package, so I worked out how I could print these from my computer and make them look semi-professional. I went to Sam's Club and picked up some small white paper bags for packaging; Jo-Ann Fabrics for some rubber stamps and ink; and Walmart for some natural jute twine to tie it all up with. I already owned all the baking supplies I thought I'd need.

I baked and delivered those first six loaves, and week after week, my orders grew. I think I finally realized the Local Loaf was succeeding when I had baked 24 loaves of bread and bags and bags of rosemary rolls the day before Thanksgiving — all for my original group of customers. Customers told their friends, and, as new employees started with the company, they'd ask to be put on "the list." My customer list continues to grow; it now hovers at well over 100 people. I bake once a week — on Thursdays for Friday delivery — and have added all manner of other baked goods, including cookies, muffins, cupcakes, and scones, plus byproducts like croutons, to my offerings.

The Local Loaf provides me with the opportunity to make income from home — no commute, no expensive work clothes or handbags, no take-out lunches, no office politics — while also providing me with a creative outlet for my baking. It also enables social interaction with a core group of customers who count on my baked goods every week. My bread business has landed me unexpected benefits as well, such as paid bread-baking demonstrations at national fairs, retail stores, and even an (unpaid) gig at the Kansas State Fair as a food judge!

make your own soap or sew your own drapes, maybe one of those skills could be developed into a marketable business. Homesteaders are remarkably talented, handy people; people who aren't, or who don't have the time (or interest) to do things for themselves, will spend money or barter to obtain a piece of your handmade ingenuity. If your product is high quality, reliable, and packaged well, chances are that folks will line up to get it.

One of the most important factors in running any successful home-based business is organization. Without the structure of a boss, fellow employees, an office schedule, and a time clock, an unorganized individual will soon falter. You must be self motivated, detail oriented and have the ability to stay organized. Keeping track of necessary inventory and projecting needs far enough in advance is important because many special items (like packaging) need to be acquired online. Running a business from home requires you to make optimal use of your time because you'll need to constantly juggle multiple priorities (business and personal) — and there are no paid vacations or holidays!

When it comes to home-based businesses, most of our experience is with food (specifically, baked goods), so this chapter will be centered on that. However, it isn't hard to extrapolate our experience with food production to other areas, especially those that rely on some aspect of the farm or homestead for success.

Why Baking?

In this day and age of busy-ness, home baking has utterly gone by the wayside. Boxed mixes are now the norm, and the concept of baking from scratch is not even in many folks' realm of consciousness. Many competent home cooks shy away from baking, believing it's a complicated science (well, it is) that it requires lots of practice to get good results (well, it does). If you already bake from scratch, this could be your opportunity. Your customers will be those people who desire a high-quality, home-baked product, yet are unable or unwilling to devote the time to baking it themselves. Simple enough. These people include working women, women who don't like to cook, women involved in multiple community events who don't have ample kitchen time; men with working wives; those who crave homemade meals, and foodies from all walks of life.

The tenet of your business should be quality. That high-quality, homemade product can only be obtained from ... you got it ... a home — not a factory. Make a serious commitment to handling each order individually, and baking everything from scratch with only the best ingredients. Boxed mixes are convenient, sure, but they cannot compare to the genuine article. Same goes for store-bought and frozen cakes, cookies, and bread — they're made in factories by machines with artificial flavorings, fillers, and preservatives in order to withstand days and weeks on the shelves. When your customers taste your fresh, unadulterated home-baked goodies, they'll know the difference and will pay to taste it week after week. Most people are gourmets at heart when it comes to baked goods — they reminisce fondly of "how grandma used to make it" and such, and they will develop a similar fondness for your products.

Working from the Home Kitchen

Having a suitable home kitchen for baking is one of the first considerations when starting a food business. Even if your kitchen isn't entirely up to the task of high-volume baking, you can certainly begin your business before doing any remodeling. Surely you already have hot and cold running water, basic appliances, and sound surfaces. Some of the most successful food businesses began in the humblest of surroundings. Margaret Rudkin, founder of Pepperidge Farm, began just this way in her home kitchen in the late 1930s.

Make sure to investigate all your state and county regulations for home-based food businesses *before* remodeling your kitchen — some states don't allow a kitchen to be used for dual purposes, so you would need a second kitchen.

Cleanliness Is Next to Godliness

When operating a home-based food business, your kitchen is the face of that business. Would you show up at the office with uncombed hair, last night's makeup, and wrinkled clothing? Not if you wanted to remain employed. Keep in mind that the kitchen should be neat and tidy all the time, not just when you are baking for customers. Wash and put away dirty dishes promptly, change dish towels and sponges often, wipe down surfaces several times a day, scrub the sink with cleanser, and sweep and mop the floor regularly. When baking, tie your hair up or back in a bandanna, put on a clean apron, and make sure your hands and nails are scrubbed clean. If customers will be picking up their orders, all of this is even more important. Food can seem quite unappealing in the midst of a dirty kitchen. Remove all suspicion of uncleanliness by keeping the kitchen very clean. If you live in a state or county that requires health department inspection, make sure you're aware of the requirements so you don't get caught unprepared.

Storage and Work Space

It's important to store your ingredients properly. Never store food in unsanitary conditions or places (like a bathroom), near moisture, or directly on the floor — that's a health code violation just about everywhere. It's ideal to have a dedicated storage space for all your baking ingredients so that when you're ready, you only have to look in one place; it also makes it easier to take inventory. If dedicated storage isn't possible, label everything you use for your business and keep it one area of the pantry. Store sacks of white flour in rodent- and pantry moth-proof containers; we use a new, small metal garbage can with a tight-fitting lid. Always store whole wheat and other whole grain flours in the refrigerator or freezer if they won't be used up within a few weeks. Whole grain flours turn rancid more quickly than refined white flours.

Work space is critical. If you don't have cleared, clean spaces on which to work, you'll soon become frustrated. Take over the kitchen island and counters for food preparation and clear off the dining room table for the cooling and packaging station. If you need more space than you have, get out a long folding table to handle the overflow. Take the time to clear everything off your work surfaces, wipe the areas down, and commence baking with a clean slate every week.

Food Safety and Prep

Some states and counties require that you take a food-handling certification course. If you've never worked in food service, do some research to make sure you understand food safety and always follow best practices — that means no dipping your finger into the batter; that's fine when preparing food for yourself, but not when operating a food business. All it takes is one sick customer to put you out of business, so remain vigilant all the time. If you suspect — even remotely — that milk, an egg, or any other ingredient has gone bad, err on the side of caution and discard it (or feed it to the pigs or compost pile). Again, it's fine to take such risks when you're cooking only for yourself, but not when preparing food for others.

Product Offerings

When planning your product offerings, begin with what you know. What are you particularly good at making? What have people raved about? Start there, then think strategically about what products might be needed by your target market. Is there a shortage or absence of good bakeries in your area? Perhaps artisan-style bread would be welcomed. What are your regional specialties and favorites? Offer products that people want (sugar cookies at Christmas and Valentine's Day, rosemary rolls at Thanksgiving and Christmas) but seem too complicated or time consuming for them to make themselves. In the beginning, limit your product offerings to just one or a few items that you make well. This will give you time to learn the business (inventory control, packaging, bookkeeping) and gauge feedback before getting overwhelmed or heading off in the wrong direction.

Before deciding on what type of food to offer, research the market, even if just casually. Peruse specialty food offerings at grocery stores, farmers' markets, health food stores, and the like. Examine pricing and packaging. Purchase anything you want to emulate or learn from. Prepare a few of your specialties and ask testers for feedback — seek those who will be honest with you rather than telling you what you want to hear.

Evaluate recipes with an eye on two factors: price of ingredients, and ease of preparation. Recipes that call for exotic ingredients that are hard to obtain in your area will create an inventory problem; expensive ingredients (like whole vanilla beans or pine nuts) won't bring a return on your investment. Focus on offering items that aren't too fussy, that can be packaged easily, and aren't too expensive to make. For example, think about making a carrot cake. Making a carrot cake involves peeling and shredding carrots; vast and varied ingredients, from dairy products to spices and nuts to raisins; baking and assembling a layer cake; and frosting the cake. A moment's thought will convince you that carrot cake would be a packaging and delivery nightmare. Also, consider whether you could ever charge enough to cover your time and the resources you invested. Instead, look around for goods that present

Good Recipe, Bad Recipe

Put on your thinking cap when evaluating recipes for possible production. Below are two recipes for chocolate chip cookies. Recipe 1 is the one I'd been using for my Local Loaf cookies for two years. I spotted another recipe that I thought might be an improvement. But upon closer inspection, I rejected it. It required more ingredients, more labor, more money. I might use the recipe when I bake for my family, but it's not suited for "production line" at the Local Loaf.

Recipe 1: Chocolate Chip Cookies

Makes 36 large cookies

4¼ cups all-purpose flour
1 teaspoon baking soda
1 teaspoon salt
3 sticks unsalted butter
2 cups dark brown sugar
1 cup granulated sugar
2 eggs plus 2 yolks
4 teaspoons vanilla
3 cups semisweet chocolate chips

Recipe 2: Chocolate Chip Cookies

Makes 36 large cookies

4¾ cups all-purpose flour (MORE)
1½ teaspoons baking soda (MORE)
2 teaspoons salt (MORE)
1½ teaspoons baking powder (MORE)
3 sticks unsalted butter
2¼ cups light brown sugar (MORE)
1¼ cup, plus 2½ tablespoons granulated sugar (MORE)
3 eggs (LESS)
2 teaspoons vanilla (LESS)
1⅔ pounds bittersweet chocolate, chopped (LOTS MORE, PLUS CHOPPING, PLUS SOLID BLOCKS OF BITTERSWEET CHOCOLATE ARE EXPENSIVE)

The Perfect Product

When I started the Local Loaf and decided to offer artisan bread, I decided on no-knead bread, baked in a cast-iron pot. The dough is mixed 18 hours prior to baking and then sits in a covered bowl, fermenting away, until 1 hour prior to baking; at that point, it's turned out, covered with a towel, then baked. My oven accommodates three pots at once, so I used three loaves as one "batch" to figure out a timetable. From time to time, customers ask me to make "sandwich bread" (in other words, kneaded, soft, loaf bread). But that kind of bread is extremely fussy — with its kneading, multiple rises, and sensitivity to ambient temperatures — so I decided against offering it. It would bring no return on my resource investment. I make kneaded bread for my family, but trying to manage 12 loaves at once, all at different stages of rise, would surely make me lose my sanity. Instead, I continually develop new no-knead-style recipes so I can offer customers something new.

fewer potential problems but are still something special. You could focus on high-quality muffins with the best ingredients, cookies, or cupcakes — all of which are less work and easy to package and deliver. Avoid offering products that require refrigeration, especially those that would take up lots of space — there are only so many cakes and goodies that can be stored in a household unit.

Ad hoc Business Plan

A formal business plan complete with executive summary, market research, and financial projections is really only necessary if you're seeking financing. You certainly do not have to write out a business plan before getting started; however, you might find an ad hoc business plan helpful in the beginning. As you gain more experience and build your business, you will probably want to do a more formal plan.

A business plan is helpful in that it articulates your daydream and lays out a vision for the future. It's the place to define your goals and the steps necessary to achieve them. Just by putting your ideas and research findings down on paper you'll gain a better understanding of the reality of running a perishable food business.

What should be included? For an ad hoc business plan, consider the following:

- **Statement of Purpose**
  Also known as the *mission statement,* your statement of purpose explains your business in a few words or sentences. Example: *The Local Loaf provides homemade, baked-to-order artisan bread made from all-natural ingredients to the local community.*
- **Description of the Business and Products**
  This should say why you chose the name for your business and why it appeals to customers. Discuss the products you offer, plan to offer in the future, and why, on both counts.

Detail delivery method, target customers, and why your products will fill a niche.

- **Values**
  This encapsulates how you want your company to be viewed by the public. Do you have a theme for your company, such as "local food," "homemade," or "all natural"? Describe your product line and the overall idea behind it.
- **Market Research**
  Describe your target market and the market research you've done to determine your product offerings. Also include Internet research on trends, and rationale for why your product is needed by your target market. Consult trade organizations, local business groups, and include any informal polls you've conducted. Don't forget a word or two about your competition — who they are and what products they sell. What makes your product superior?
- **Promotional Strategy**
  Detail how you will market your business and how you will cultivate a customer base. Describe promotional activities like Facebook and Twitter, creating flyers and brochures, appearing on radio or television shows, or participating in craft or food fairs.

Making It Legal

Of all aspects of running a home-based food business, this is the least fun. However, by formalizing your business and making it legal, it will open doors for you and allow your mind to wander to more creative aspects of the business.

- **Step 1: File**
  Begin by creating either a sole proprietorship or a partnership. Both have pros and cons, but know that a sole proprietorship is the simplest form of business; the owner declares all profits on his or her income tax return. Depending on where you live, you may need to obtain a business license or pay special fees.

  If you're starting a joint venture with your spouse or friend, you will have a partnership. As close as you may be with any potential partner, things can go south really quickly in business. Prepare a written legal agreement that spells out how much time, money, and work each partner will contribute; it should detail, among other things, division of labor, resolving disputes, and dissolution of the partnership. As with a sole proprietorship, you may need to obtain a business license and pay special fees. Consult the IRS for more information.
- **Step 2: Register Your Business Name**
  Operating under any name that is not your legal name requires that you register it as a fictitious business name (otherwise known as a *DBA* — doing business as). Obtain a DBA form from your local government office or

office supply store and register it at the courthouse. Filing this form will ensure that no one else in your county is using the same name, and, once you file, no one else can use it in the future. A DBA will be required to open a bank account with your business name, which you'll need to deposit checks made out to the business.

- **Step 3: EIN (Employer Identification Number)**
  If your business is a sole proprietorship, you can use your social security number for tax purposes. If you created a partnership, you'll need to get an EIN so the IRS can identify your business and track your taxes.
- **Step 4: Business License**
  Contact your county, city, or state government offices to inquire about the need for a business license. This is separate from a food production license. You may also need a business license to open a bank account with your business name.
- **Step 5: Food Production License**
  When it comes to food processing regulations, every county and state is different, with varying levels of restrictions. In heavily regulated locales, a home-based food business may be out of the question. In others, it may be perfectly legal even without a health permit or food-handling license. Find out what the codes and laws are in your area and which agency does the licensing. Start by calling your local health department and ask for the guidelines on home-kitchen processing (you might even find these online). Licensing will depend on the type of food handling you'll be doing. In Kansas, for example, baked goods are considered a "low risk" food so virtually no regulations exist for home processing. However, selling items like meat and cheese require state licensing. In some areas, it may depend on how much income you generate from the business. Contact your local business authorities to find out what the rules are. Adhere to them.
- **Step 6: Insurance**
  Homeowners' insurance will likely cover a home-based food business, but you will need to purchase an additional rider to receive coverage. Commercial insurance (also known as a vendor policy) must be purchased to cover you for product liability. Check with your insurance company.

Equipment and Packaging

The equipment you'll need may already be in your kitchen. Don't go out and purchase all-new machines and tools before you have the customers and cash flow to justify them. However, if there are a few key items, like a stand mixer or handheld electric mixer that would make the business venture more efficient, by all means purchase them; they can be itemized on your tax return as business expenses.

Packaging is key to your business success. Without professional-looking packaging, your product will look like a child's bake-sale offering. Forget plastic baggies and curly ribbon — think up something unique and appealing. For Local Loaf products, we use plain paper bags (white paper for bread, and brown Kraft paper for baguettes and some baked goods) wrapped up with jute twine, which lends a natural look. Using rubber stamps and brown ink, the name of the bread variety is stamped on the outside of the bag. Clear cellophane food bags are used for cookies, and brown Kraft bakery boxes (made with recycled paper) with windows are used for muffins, cupcakes, and delicate cookies. A Local Loaf tag is affixed to all items. The tag has a professional business card on one side and the ingredient list on the other (printed on our home computer), which satisfies the state's requirements for selling baked goods.

The cost of packaging is an eye-opening line item when it comes to profit and loss. If you limit yourself to buying what's locally available at craft stores like Michael's, you'll be paying retail prices for boxes and bags (usually in the $1 range for a bakery box); that's a huge chunk of your profit eaten up — unless you add that cost to your selling price.

Sourcing unique materials online from specialty packaging merchants is a better way to go. Recycled paper bakery boxes of all sizes and shapes, multicolored twine, printed muffin and cupcake liners, and cellophane food bags can all be had for a fraction of the retail price when you buy in bulk online — that $1 bakery box can be had for around 35 cents. For sources, see "Resources for Packaging Materials and Specialty Kitchen Tools on the next page."

Pricing

To determine the price for your homemade baked goods or other products, several factors must be taken into account. First will be the actual cost of

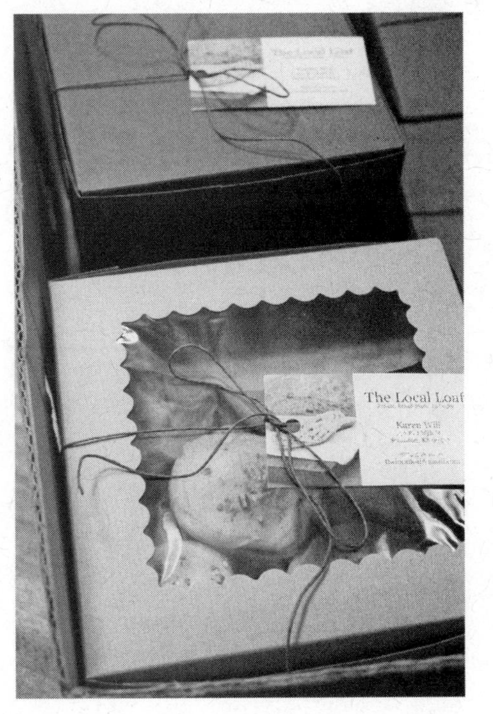

Fig. 12.2: *Packaging is key to your business's success. We use brown Kraft bakery boxes (made with recycled paper) with windows for muffins, cupcakes and delicate cookies. A Local Loaf tag is affixed to all items.*

CREDIT: KAREN K. WILL

the ingredients; next, the electricity, gas, or other power necessary to make it; then there's packaging; the value of time spent; and, finally, making an allowance for profit. Figure these things carefully before arriving at a price because customers are sensitive to price changes (particularly price hikes) and don't like it when the number is constantly fluctuating. Simply comparing prices of similar goods at the markets isn't a great gauge of what your price should be; unless you have a wholesaler's license and shop at Costco or Sam's Club, you'll be paying retail prices for ingredients — far more than a factory that is mass-producing baked goods.

Ingredient	Size/Price per package	Units per package	Price per unit
Flour (all-purpose)	5 lbs/$2.89	1 lb=.58 4 cups per lb	.15 per cup
Butter	1 lb/$3.50	2 cups	$1.75 per cup
Brown sugar	2 lbs/$2.50	4½ cups	.56 per cup
Granulated sugar	5 lbs/$3.39	1 lb=.68 2¼ cups per lb	.30 per cup
Eggs	1 dozen/$2		.17 per egg
Vanilla paste	32 oz/$40	64 tablespoons or 192 teaspoons	.63 per tablespoon .21 per teaspoon
Chocolate chips	12 oz/$3.50	2 cups	1.75 per cup

The first deciding piece of information you'll need is exactly how much the particular item costs to make. Do a cost analysis for each item by breaking down the ingredients, figuring the cost

Resources for Packaging Materials and Specialty Kitchen Tools

- MrTakeoutBags (mrtakeoutbags.com) — Food-service packaging of all kinds including recycled Kraft bakery boxes, bread bags, muffin liners
- Uline (uline.com) — bakery boxes and bags, twine
- Uprinting (uprinting.com) — online business cards, letterhead, brochures, hang tags, etc. that you design
- Kitchen Krafts (kitchenkrafts.com) — Food-crafting supplies including packaging materials and specialty tools
- Chef Depot (chefdepot.com) — Emporium of gourmet kitchen equipment like plastic dough scrapers and high-heat potholders
- Pleasant Hill Grain (pleasanthillgrain.com) — Homesteader's kitchen supply, plus hard-to-find large poly bread bags
- King Arthur (kingarthurflour.com) — specialty baking ingredients
- Modern Baking (modern-baking.com) — Industry publication that's useful for keeping up with trends
- Bread Bakers Guild of America (bbga.org) — Artisan baking community organization offering classes, resources, and more

of all other inputs, and doing some simple arithmetic as shown at right.

Sample Cost Analysis

1 Batch of Chocolate Chip Cookies (36, sold by the half dozen)

Item	Cost
• 4½ cups all-purpose flour =	.64
• 1 teaspoon baking soda =	.01
• 1 teaspoon salt =	.01
• 3 sticks (1½ cups) unsalted butter =	2.63
• 2 cups dark brown sugar =	1.12
• 1 cup granulated sugar =	.30
• 4 eggs =	.68
• 4 teaspoons vanilla =	.84
• 3 cups semisweet chocolate chips =	5.25
Total cost of recipe:	11.48
Cost of one cookie:	.32
ADD	
• Power =	.25
• Cellophane bags (6) =	.72
• Twine (6 lengths) =	.30
• Tags (6) =	.60
• Ingredient lists (6) =	.30
• Labor (1 hour at minimum wage of $7.25) =	$7.25
Total cost of extras:	9.42
Cost of one cookie:	.26
Overall total:	**$20.90**
Overall cost of one cookie:	**.58**
Retail price: $1 per cookie (6 for $6 = $36)	**$36**
Profit:	**$15.10**
Total net:	(profit + labor) **$22.35**

Marketing

Now that we've got this all figured out, the last bit is acquiring some customers. In our experience, it's not profitable (or even possible to break even) to sell products wholesale to retailers for resale. Businesses like restaurants, cafes, gourmet food stores, caterers, and the like are used to purchasing from food-service companies (like Sysco, who charges 50 cents for a loaf of bread), and their profits depend on those low costs. If you approach retail businesses and expect to get $5 for a loaf of your bread, you'll come away disappointed; there's simply "no meat left on the bone" for the merchant to make a profit. Instead, focus on direct marketing your products.

Before you start scouting for customers, you'll need some promotional materials to help spread the word.

If you don't already have some kind of desktop publishing program on your computer like Microsoft Publisher, it might be worth the investment to buy it. With such a program (plus a *Dummies* book) you can create your own flyers, brochures, business cards, and menus — all the marketing materials you'll need to promote your business. Without desktop publishing, go ahead and create simple materials in any program you know. If this isn't your forte, do what you need (recruit a friend or your child) to make sure the materials are professional looking and free of misspellings and typos. Think about the last restaurant menu you read that was full of bad grammar and typos. Did it inspire your confidence in the food?

March 25, 2010 / Volume 1, Issue 2

The Local Loaf

Artisan Bread Delivered To Your Desk

Karen Keb, baker
thelocalloaf@gmail.com

 **Easter Special: raisin cinnamon $4.50**

Essence of Time

We often hear that "time is money," but when it comes to no-knead bread, time is free and it's what gives the bread its unique flavor.

A loaf is started about 19 hours before it ever goes into the oven. It's allowed to slowly rise, letting the yeast luxuriate to life.

Because of the bread's relationship with time, I had what turned out to be a kitchen disaster last week. I bake two loaves at a time and during the second batch, there was a bizarre blast of heat in the oven. I checked it with four minutes left on the timer; went to my computer, came back and the loaves were charred black on top! I was miffed because there was absolutely nothing I could do ... I couldn't just bake another two loaves "real quick" to make up for the oven malfunction!

I guess this is where the word "artisan" comes into play. My bread isn't made in a factory, so it will never be uniform from week to week, or loaf to loaf. In a funny way, each loaf is a one-of-a-kind. Even the room temperature plays a role in how the bread turns out. (Expect higher and higher loaves this summer!)

The Barter is Back!

Do you can salsa? Maybe you bake a mean cherry pie or you're harvesting spinach from your cold frame? Or maybe you have an in-demand skill like sewing. If you'd like to propose a trade for bread, please do. If it's something I (or Hank) need, we'll make a deal!

E-mail Your Order to: thelocalloaf@gmail.com
Bread will be delivered via "Hank Express"

bread varieties available for order

White	$4
Whole Wheat	$4
Rye	$5
Raisin Cinnamon	$5
Walnut Raisin	$5.50
Irish Brown	$5.50

Next Week

- Bread will be delivered on Wednesday and Friday
- Sunday, April 4, is Easter. Still time to order!

Kindly return cloth bags to Hank at your convenience.

Due to the amazingly generous response for bread, I doubled my baking from 4 loaves per day (twice a week) to 8 (that's 16 loaves per week)! I love doing it, so continue to make your requests and I'll fit you in. I'll always let you know what day you'll get your bread, or make arrangements if you need it on a certain day.

Recipe: Bread Pudding

- 4 cups raisin walnut bread, cut into 1-½" chunks
- 1 lb. cooking apples (2 medium), peeled, and cut in bite-sized chunks
- handful of raisins
- handful of chopped walnuts
- ½ cup sugar (add more to taste)
- 3 cups milk
- 6 large eggs
- 1 T cinnamon
- dash of clove and nutmeg

Preheat oven to 350°. Put the first four ingredients in a large bowl; whisk the sugar, milk, eggs and spices together and pour over the bread cube mixture. Pour all into a buttered 11" x 17" baking pan. Cover with foil and cook for 35 minutes; remove the foil, sprinkle with a little more cinnamon, then bake for another 20 minutes, until the center of the pudding is no longer soupy. It will cook a little more as it cools down, so don't overcook it. Serve with vanilla ice cream while still warm. —courtesy K.C. Compton

New: Irish Brown

I am now offering Irish Brown bread, made with **Guinness** stout and **buttermilk**. It's delicious with a dark, crunchy crust, moist crumb, and a slightly bitter bite (thanks to the Guinness).

Fig. 12.3: *Newsletter, flyers, or brochures should contain a sentence or two about your business, descriptions of your products, and a price list. You'll also want to include your ordering and delivery protocol.* CREDIT: KAREN K. WILL

Flyers or brochures should contain a sentence or two about your business, descriptions of your products, and a price list. You'll also want to include your ordering and delivery protocol. Coupons or special offers entice hesitant customers, so think about offering an initial or weekly special that won't cut into your profits too much.

Other ideas for self promotion include offering demonstrations at fairs and retail establishments that offer cooking classes; you'll get face time with many interested people, some of whom may wind up being your customers. A simple website with your basic product offerings will legitimize your business in the eyes of potential customers, and a weekly or monthly newsletter highlighting specials and recipes can be sent out if you gather customers' email addresses.

Direct Marketing Avenues

Farmers' Market

Food vendors are always welcomed and needed at farmers' markets, street fairs, food and craft fairs, and church and school bazaars. Call the organizers and inquire about becoming a vendor.

Office Buildings

Offices — including banks, schools, doctors, and hospitals — are a great source of customers. Employees who work full time often desire — but don't have the time — to bake or cook. Offices are always needing special items for parties and lunches. Offices also provide the convenience of one location to deliver to and collect money from — your customers will be there all day, so your delivery schedule can be quite flexible. It's helpful to know someone at the particular office you're targeting who can vouch for you and your products (and, ideally, even handle the deliveries and money collection). If you don't know anyone on the inside, inquire at your bank, your child's school, beauty shop, or your doctor's office about their need or desire for your services.

Your Neighborhood

Your surrounding neighborhood, or the closest concentration of houses and apartment buildings is a good source of potential customers as well. Take your flyers or brochures to residential areas or apartment buildings and place one in each mailbox.

Groups of people who gather on a regular basis are another perfect

Home Baking Business Books

Two excellent guides with varying levels of detail are: *A Gold Mine in Your Kitchen: Home Baking for Profit* (an oldie but a goodie from 1967) by Helen Scott (out of print, but available used on Amazon), and *Start & Run a Home-Based Food Business* by Mimi Shotland Fix (2009).

Lessons Learned from The Local Loaf

Here is an inside look at what I've learned from operating my farm-based baking business, The Local Loaf, for the past three years.

1. Customers are fickle. Even if they rave about one of your regular products, they are always on the lookout for something new. So you should always offer new products — at least once every two weeks, if not every week. This means you'll always need to be on the lookout for new offerings, looking at trends, and considering seasonal offerings to capitalize on customers' desire for something new and exciting. Once I began offering new cookies, muffins, and bread varieties every week, my orders doubled. You don't have to reinvent the wheel *every* week, just rotate certain items on and off the menu, which also serves as a "get it while it's hot" incentive.
2. Be consistent and reliable — no excuses! I established from the very beginning that I would bake on Thursdays, and delivery would be on Friday morning. I try to stick to this schedule every week without fail. If you're constantly coming up with excuses as to why you can't uphold the schedule, your customers won't be able to rely on you and will find other places to spend their money. If you do have to interrupt the schedule once in a while for an out-of-town trip or emergency, just make the announcement short and to the point — no elaborate excuses or sob stories — and promise to be back next week, baking as usual.
3. Be amenable to special requests. If someone needs a special order on a non-delivery day, or someone wants an item made with a different ingredient, be open to it if you can. Being the "Bread Nazi" doesn't foster goodwill! One of my most successful cookies came from a customer request. I offered lemon-ricotta cookies one week, and a customer asked if I'd consider making orange-ricotta cookies, with one half dipped in chocolate. After quickly considering it, I said yes and executed the same recipe, just substituting orange juice and orange zest for the lemon. ☞

Fig. 12.4: *You don't have to reinvent the wheel every week, just rotate certain items on and off the menu, which also serves as a "get it while it's hot" incentive.* Credit: Karen K. Will

I melted bittersweet chocolate chips and dipped half the cookie in it. I have yet to have more orders for a cookie.

4. It's key to have an inside person to help you. About 75 percent of my business is concentrated in one office building. My husband's assistant loves to help with the bread deliveries and money collection, and she is very good at it. For some reason, no one begs off paying her until next week (which used to happen when Hank was handling it). Having a third party handle the delivery and collection makes the whole process smooth and efficient for everyone involved. And you don't have to go around with your hand out after baking all morning. Although she never asks for any compensation in return for her services, I always give her something as payment — a loaf of bread or a few cookies to thank her for her time and effort. If you find someone willing to help you out in this manner, you could offer products or a small percentage of the take as compensation.
5. Thank your customers and give back occasionally. Think about what you expect in terms of customer service at the places you patronize and go beyond that. Always keep a positive tone in your communications. Thank each customer for their business. I once had a customer place a very large special order for a non-delivery day (Monday, which meant all-day baking on Sunday) who then cancelled Monday morning due to a "family emergency." Instead of suspecting impropriety, I simply wished her the best outcome for her emergency, then turned her large order into "Customer Appreciation Day" for the office. An email was sent out to everyone on my regular list to "come and get it." It was a nice way to turn lemons into lemonade. I also donate food to charity picnics and the like, as a way of giving back.

customer base. Look in local papers for community groups (master gardeners, Toastmasters, church groups, etc.) and find out when they meet. As with office clients, having everyone in one location means easy delivery and collection. Plus, folks who participate in community groups are often busy and socially connected. You can ask them to refer friends and associates to your business.

Kitchen Sales

If you're comfortable with customers coming into your home, you could offer the option of pick up. If your customer base is comprised of friends and acquaintances, this may be perfectly safe; however, we don't recommend opening up your home to complete strangers, especially if you're home alone during pickup times.

Managing Finances

For tax purposes, it's important to keep a log of all your business purchases and sales. A simple ledger book or lined notebook will suffice, or a spreadsheet on your computer will do (make sure to back it up in case of a computer meltdown). Record weekly sales figures, and keep receipts for all ingredients and equipment purchased for the business. Depending on how analytical you want to get, you could keep multiple spreadsheets to track profit and loss, expenses, popularity of particular items, weekly sales by year to determine busy periods, etc. It's all up to you!

The Question of Expansion

You'll find that after realizing a little business success, people will ask you if you plan to open a restaurant, café, or bakery. If that's your long-term goal, then you're already on your way with a customer base. But know that operating a bricks-and-mortar establishment is an entirely different proposition than a home-based food business. Overhead costs, employees, round-the-clock food preparation, vendors, insurance, commuting, etc., can all conspire to take the fun (and profit) out of this job. It's OK to stick to your original plan of working from home. There's no reason you should feel aimless if expansion isn't part of your plan. The American economic system dictates that a business must grow every year to be healthy, but if you're successful by your own definition, be content. Listen to what your gut tells you about expanding your business.

Other Home-based Business Ideas

Activities you're already engaged in are a natural fit, but anything that relies on some aspect of the farm or homestead would also be a natural choice. Consider:

- Meat and eggs — farmers' market or farm sales
- Garden crops — roadside stand, farmers' market, CSA
- Yarn, fiber, sewing — direct market through Internet, craft and livestock shows, or CSA
- Homemade soap, detergent, fabric softener, deodorant, other body products — farm sales, farmers' market, Etsy
- Homemade beeswax (or other) candles — farmers' market, Etsy
- Honey — back door sales
- Crafts — Etsy
- Rustic or salvaged furniture or antiques — Craigslist, eBay, flea market or local antique shop booth space
- Writing, blogging, photography
- Dried herbs, spices, teas
- Teaching classes on homesteading skills like canning, baking, woodworking, fence building, livestock care, gardening, etc.

Cultivate Your Clients

Our good friend, Sandra Johnson, of Baldwin City, Kansas, was responsible for turning me onto homemade soap, including laundry detergent and fabric softener. She's made soap for decades, and, as she's a very conscientious teacher, she held my hand through the entire process, demystifying it for me.

Sandra used to sell soap at craft fairs and holiday bazaars, but tiring of the unpredictable sales and the whirlwind of preparation, she brought the business back home and cultivated a select group of clients who pre-order her soap and pay top dollar for it. I'm convinced this is the best business model for a homestead food or craft business: Find your customers, wherever they may be, and cater to those who want to pre-order and pay for your products. After all, you have a limited amount of time and energy, so why not capitalize on that fact? Small-batch, handcrafted items are unique — customers can only get your product from you (and only for a limited time).

Think hard before spending endless money on a booth or storefront somewhere, making tons of product on speculation and hoping someone comes along and buys it. That's wasteful of precious energy and supplies. Nothing will dash a fledgling home-businessperson's hopes faster than the sound of crickets chirping at their craft fair booth.

Fig. 12.5: Credit: Karen K. Will

Afterword

We've both authored books before, but when we decided to do one together, we had to think long and hard about what to cover. When it came down to it, we decided to go with topics that we felt would be of use to readers — things that modern homesteaders like us could put into practice — but also things that we held close to our hearts. For Hank, this included everything related to raising or building what is needed instead of purchasing similar (yet inferior) products from big, faceless corporations. For Karen, it was everything relating to cooking and handcrafting that could provide necessities for home and community — and finding ways to rely on other handcrafters for most everything else. The projects you've read about here are some of our favorites. With each one completed, we've been inspired with even more ideas and ways to invest our time similarly.

Along the way, there were so many projects — like building a just-right-sized portable feed bin for the hogs and chickens, or installing a key cabinet to hide some old water spigots — that made us turn to each other and say, "Oh, we've got to include this in the book!" But, there just wasn't enough space or time. On the horizon next is the building of an old-fashioned shepherd's hut (sometimes known as "gypsy caravan") to move around our pastures so we can enjoy time with our sheep and use it as a bit of a retreat, while still remaining on our farm. Hank has already sourced the running gear on Craigslist; it's parked out back of the barn waiting for its next life to take shape. We got the idea for it after seeing an article in the British *Country Homes & Interiors* magazine about an English couple in Dorset who have revived this old-time craft and have now turned it into a thriving business, Plankbridge Shepherd's Huts.

Our lifestyle allows us plenty of time to reflect; we've come to believe that the best part about living in wide open spaces is that we have the freedom to craft the life we desire without too many folks infringing on our dreams. With very little money and a blatant disregard for "what people think," we make our own fun, and you can too.

Hank & Karen Will
Osage County, Kansas
September 2012

Photo Credits

Section 1:

Section 1, page 3: All photos courtesy of Karen K. Will

Chapter 1, page 5: Photo courtesy of Karen K. Will

Chapter 2, page 31: Photo courtesy of Karen K. Will

Chapter 3, page 47: Photo courtesy of Karen K. Will

Section 2:

Section 2, page 63: Photo credits left to right courtesy of Karen K. Will, Oscar H. Will III, Wendy Slatt, Karen K. Will.

Chapter 4, page 65: Photo courtesy of Karen K. Will

Chapter 5, page 87: Photo courtesy of Karen K. Will

Chapter 6, page 107: Photo courtesy of Karen K. Will

Section 3:

Section 3, page 123: Photo credits left to right courtesy of Oscar H. Will III, Oscar H. Will III, Oscar H. Will III, USDA

Chapter 7, page 125: Photo courtesy of Karen K. Will

Chapter 8, page 139: Photo courtesy of Oscar H. Will III

Chapter 9, page 151: Photo courtesy of Karen K. Will

Section 4:

Section 4, page 165: Photo credits left to right courtesy of Karen K. Will, Karen K. Will, Oscar H. Will III, Karen K. Will

Chapter 10, page 167: Photo courtesy of Marjorie Keb

Chapter 11, page 199: Photo courtesy of Oscar H. Will III

Index

About the Authors

Oscar H. (Hank) Will III is a farmer, scientist and author, known for seeking and implementing creative farmstead solutions. The editor-in-chief of *GRIT* magazine, Hank has published hundreds of articles and five books on a range of topics including antique farm machinery.

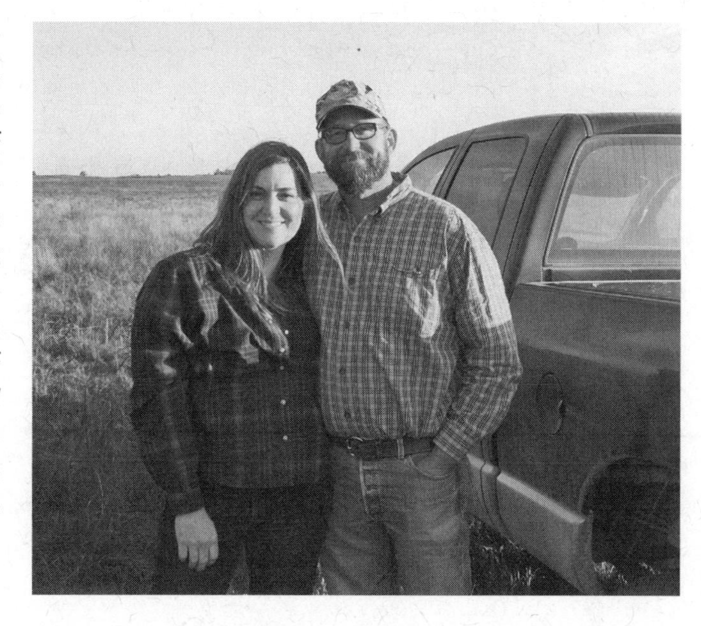

Karen K. Will is editor of *Heirloom Gardener* magazine and author of *Cooking with Heirlooms: Seasonal Recipes with Heritage-Variety Vegetables and Fruits*. She operates Prairie Turnip Farm with her husband Oscar H. Will III.

New Society Publishers

ENVIRONMENTAL BENEFITS STATEMENT

New Society Publishers has chosen to produce this book on recycled paper made with **100% post consumer waste,** processed chlorine free, and old growth free.

For every 5,000 books printed, New Society saves the following resources:[1]

35	Trees
3,189	Pounds of Solid Waste
3,508	Gallons of Water
4,576	Kilowatt Hours of Electricity
5,797	Pounds of Greenhouse Gases
25	Pounds of HAPs, VOCs, and AOX Combined
9	Cubic Yards of Landfill Space

[1]Environmental benefits are calculated based on research done by the Environmental Defense Fund and other members of the Paper Task Force who study the environmental impacts of the paper industry.

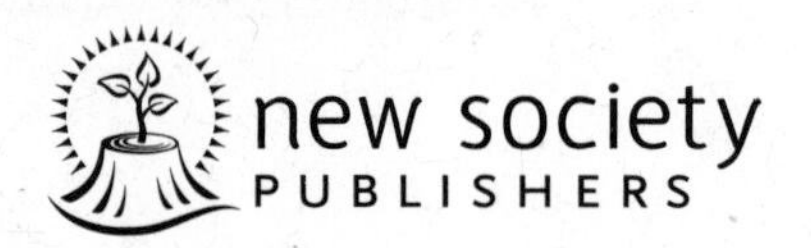